Communicating Climate Change in Russia

The attitude of Russia towards climate change is extremely important for the success of climate change control policies worldwide, as Russia, with its cold climate and vast resources of carbon fuels, is one of the world's biggest polluters. Moreover, Russia frequently comes across as not being very interested in containing environmental pollution. This book explores how issues to do with climate change are handled by the Russian media. It discusses how the state and economic elites have influenced Russia's environmental communication, with the state's control of the media strengthening since Putin came to power, and with control being exercised in some cases by ignoring or silencing the key issues. However, the book also shows how, recently, elites and the state in Russia have begun to realise that it is in the state's best interest to pursue more climate-oriented policies. The book concludes by examining how the communication of climate change issues in Russia could be improved and by assessing the extent to which a recent change in state climate policy could mean that media coverage of climate change in Russia will keep increasing.

Marianna Poberezhskaya is a Lecturer in International Relations at Nottingham Trent University, UK.

BASEES/Routledge Series on Russian and East European Studies

Series editor: Richard Sakwa
Department of Politics and International Relations, University of Kent

Editorial Committee:

Roy Allison, St Antony's College, Oxford

Birgit Beumers, Department of Theatre, Film and Television Studies, University of Aberystwyth

Richard Connolly, Centre for Russian and East European Studies, University of Birmingham

Terry Cox, Department of Central and East European Studies, University of Glasgow

Peter Duncan, School of Slavonic and East European Studies, University College London

Zoe Knox, School of History, University of Leicester

Rosalind Marsh, Department of European Studies and Modern Languages, University of Bath

David Moon, Department of History, University of York

Hilary Pilkington, Department of Sociology, University of Manchester

Graham Timmins, Department of Politics, University of Birmingham

Stephen White, Department of Politics, University of Glasgow

Founding Editorial Committee Member:

George Blazyca, Centre for Contemporary European Studies, University of Paisley

This series is published on behalf of BASEES (the British Association for Slavonic and East European Studies). The series comprises original, high-quality, research-level work by both new and established scholars on all aspects of Russian, Soviet, post-Soviet and East European Studies in humanities and social science subjects.

Communicating Climate Change in Russia

State and propaganda

Marianna Poberezhskaya

Routledge
Taylor & Francis Group

LONDON AND NEW YORK

First published 2016
by Routledge
2 Park Square, Milton Park, Abingdon, Oxon OX14 4RN

and by Routledge
711 Third Avenue, New York, NY 10017

First issued in paperback 2017

*Routledge is an imprint of the Taylor & Francis Group,
an informa business*

British Library Cataloguing in Publication Data
A catalogue record for this book is available from the British Library

Library of Congress Cataloging-in-Publication Data
A catalog record for this book has been requested

ISBN 13: 978-0-8153-5502-1 (pbk)
ISBN 13: 978-1-138-83227-5 (hbk)

Typeset in Times New Roman
by Apex CoVantage, LLC

To my family.

Dedicated to the memory of Boris Poberezhskiy (1973–2014)

Contents

Illustrations

Figures

Tables

Abbreviations

BRICS	Brazil, Russia, India, China, South Africa
CDA	Critical discourse analysis
CO_2	Carbon dioxide
COP	Conference of the Parties under the United Nations Framework Convention on Climate Change
CPRF	Communist Party of the Russian Federation
CPSU	Communist Party of the Soviet Union
EIA	US Energy Information Administration
EU	European Union
FOM	Public Opinion Foundation (*Fond Obshchestvennoe mnenie*)
G8	Group of Eight: Canada, France, Russia, USA, UK, Germany, Italy, Japan
GDP	Gross domestic product
GHG	Greenhouse gas
Goskompriroda (or Goskomekologiya)	Russian State Committee on Environmental Protection
IPCC	Intergovernmental Panel on Climate Change
JI	Joint Implementation
KP	*Komsomol'skaya pravda*
NGO	Non-governmental organisation
OAO	Open joint stock corporation (*Otkrytoe aktsionernoe obschestvo*)
OPEC	Organization of Petroleum Exporting Countries
PrM	Propaganda Model
RF	Russian Federation
RG	*Rossiyskaya gazeta*
Roshydromet	Federal Service for Hydrometeorology and Environmental Monitoring
RSFSR	Russian Soviet Federative Socialist Republic
RZhD	Russian Railways (*Rossiyskie zheleznye dorogi*)
SR	*Sovetskaya Rossiya*
UNESCO	United Nations Educational, Scientific and Cultural Organization
UNFCCC	United Nations Framework Convention on Climate Change

USAID	United States Agency for International Development
USSR	Union of Soviet Socialist Republics
VTsIOM	Russian Public Opinion Research Centre (*Vserossiyskiy Tsentr Izucheniya Obshchestvennogo Mneniya*)
WTO	World Trade Organization
WWF	World Wildlife Fund

Note on transliteration

In this book, I use a Library of Congress transliteration system for Russian language. Well-known names appear in their most common transliterated form (for example, Bedritsky instead of Bedritskiy).

Acknowledgements

The idea for this book was inspired by my interest in the way Russian media operate. Working as a journalist in my Siberian hometown for a few years gave me a taste of what the system is like from inside and made me curious to find out more about the impact it has on the 'outsiders' and what do they make of it. It also made me wonder whether the system should ever change or evolve into something else. My interest in the topic of climate change was discovered at the very start of my academic career during the pursuit of knowledge of what global challenges our world is facing. This research project was mostly conducted during my doctoral studies at the University of Nottingham (UK) and its completion would not have been possible without the rigorous supervision of Dr. Matthew Rendall and Prof. Mathew Humphrey. Thus, I would like to thank them for their insightful guidance and outstanding academic support. I also would like to acknowledge Dr. Bettina Renz, Dr. Neil Gavin, Prof. Cees van der Eijk, Dr. Pauline Eadie, Dr. Lucy Sargisson, Dr. Mark Wenman, Dr Nataliya Danilova, Dr Olga Khrushcheva, Dr Maria Zurnic, as well as two anonymous reviewers for their valuable comments on individual chapters. In particular, I want to thank my husband Dr Imad El-Anis and my parents Olga Poberezhskaya and Grigory Poberezhskiy for their constant and unconditional support. Lastly, I would also like to express sincere appreciation to all of the interviewees who were willing to share their professional experience and expertise with me. All of your contributions were invaluable to this study, but of course any mistakes, omissions or misinterpretations are very much my own.

Material from chapters 1, 4 and 5 first appeared in Poberezhskaya, M. (2014) 'Media Coverage of Climate Change in Russia: governmental bias and climate silence', reprinted with permission from *Public Understanding of Science* (doi: 10.1177/0963662513517848): 1–16.

Introduction

> Change will come about only if people understand the scientific realities of why we need to fight climate change. If you don't get that message clearly, then obviously you are not going to see any changes whatsoever. – P. K. Pachauri, chairman, Intergovernmental Panel on Climate Change (IPCC)
>
> (UNESCO 2009)

In 2003 President Vladimir Putin made his infamous joke at the International Conference on Climate Change in Moscow that 'an increase of two or three degrees wouldn't be so bad for a northern country like Russia. We could spend less on fur coats, and the grain harvest would go up' (Pearce 2003). Taking into consideration the history of Russian mass media and the state's influence over it, this statement could have potentially meant that one of the most intriguing issues of our time is being dismissed and ridiculed in one of the world's biggest polluters – Russia. However, as this book demonstrates, the problem of communicating climate change in Russia is not that simple.

Russia provides a fascinating case for studying climate change communication. The Russian mass media system went through 70 years of state control and some argue that this Soviet legacy can still be observed (Oates 2007). Then after the collapse of the Soviet Union, mass media gained significant freedom and power that to some extent influenced the political events of the turbulent post-*perestroika* years (Zassoursky 2004). Eventually, with the introduction of new political and economic regimes, Russian mass media had to adjust to the realities of the free market economy which their Western colleagues had been dealing with for many decades. The introduction of the free market was not the last modification Russian mass media have had to deal with. When Vladimir Putin came to power in 2000 and started to implement his policy of strengthening and centralising state power, media once again had to adjust accordingly. As a result the Russian mass media system became a unique hybrid of the 'fourth estate' (the fourth branch of power next to the judiciary, the executive and the legislature),[1] which tries to manage its way through the free market economy and at the same time has to cope with the restraints imposed by the state.

Another valuable contribution of this book comes from the significance of Russia in the world's climate change policy. Russia is one of the world's largest producers of greenhouse gases (GHGs) (Doyle 2009), mainly from extracting and burning fossil fuels. 'Russia holds the world's largest natural gas reserves, the second largest coal reserves, and the ninth largest crude oil reserves' (EIA 2013), which play a key role in the state's economy. For example, oil and gas exports together are responsible for around 15 percent of overall Russian GDP (Tekin and Williams 2009, p. 340). Furthermore, due to the vast territory, severe weather conditions and carbon-intensive nature of the economy, Russia relies heavily on fossil fuels for domestic consumption (Perelet et al. 2007). The priority of economic development in Russia for decades moved environmental issues to the background of political discourse and made it very unlikely that Russia would purposefully commit to a reduction of its economic carbon dependency in order to 'save the world' from climate change (which is still relevant). On the other hand, after the split of the USSR, Russia experienced an involuntary drop in GHG emissions. So in comparison to the late 1980s when Soviet industrial production was high, in the early 1990s due to economic collapse, Russia's GHG emissions were cut tremendously. Eventually this fact gave Russia significant bargaining power during the major international negotiations on climate change, which arguably Russia has used and abused, to the extent that it was even accused of 'environmental blackmail' (Henry and Douhovnikoff 2008, p. 451).

Recently, the situation has started to change and climate change is more often perceived as a policy of 'opportunities' rather than a policy of 'costs' (Giddens 2010). The evolution of the perception of climate change risks can be seen through the words of the country's leaders in different decades, from Putin's famous remark about less money spent on fur coats (Muenchmeyer 2008) to Medvedev's recognition of climate change's anthropogenic character and its threat to Russia's security (President of Russia website 2010). More often experts and state leaders started to talk about Russia's vulnerability to climate change consequences rather than its questionable benefits, by bringing attention to the fact that climate change is happening faster in Russia (Charap 2010) and has provoked weather abnormality that causes severe economic losses. Furthermore, former President Medvedev's statement at the Copenhagen Conference in December 2009, accepting the Climate Doctrine, the appointment of a president's adviser on climate change, as well as the realisation of Russia's great potential for the de-carbonisation of its economy through the development and implementation of steps for energy efficiency (which becomes profitable for the country) – arguably all can serve as evidence of alterations in the state's rhetoric on climate change and policy reorientation towards 'climate pragmatism'.

Overall, this book investigates a number of equally important issues:

- Firstly, it looks closely at the Russian media system and how the years of regime change have influenced it and how it has adapted to the free market economy.

- Secondly, this book explores how climate change is communicated in Russia and what factors or actors are involved in this process. It should be noted that this study contributes to a rapidly growing body of literature on media communication of environmental risks by adding an analysis of a state which to date has mostly been overlooked. Apart from a limited number of studies (Tynkkynen 2010, Wilson Rowe 2009, Yagodin 2010), Russia is rarely mentioned in this regard.
- Thirdly, in order to test the assumption of the influence of state policy, this study explores Russia's state policy on climate change (which can be seen as extremely questionable and ambiguous, but also vital to the world's climate change mitigation strategy).
- Finally, getting all of these jigsaw pieces together, this book answers the following question: If state policy is indeed an independent variable in the media coverage of climate change in Russia, then what can we expect from Russia in terms of communicating one of the most important environmental issues – climate change? Carvalho (2008, p. 164) claims that 'understanding the evolution of matters such as war, terrorism or climate change, and the ways they are interdependent in relation to the media, is one of the most important contributions to be made by social researchers'. I argue that in the Russian case it is so important because media coverage does not only demonstrate what the audience learns about climate change but also how the state approaches the problem. As Russia plays a great role in the climate change mitigation process and, due to its natural resources, has a large capacity to influence climate change in one way or another, to study the media discourse on climate change in Russia is in itself a great step in the development of this area of study.

Methodological considerations

The methods used in this research project for data collection and analysis can be broadly divided into two categories: media analysis and elite interviews.

Media analysis

A detailed explanation of the media analysis methods is presented in Chapter 4; however, some key points are outlined here. This project uses both qualitative and quantitative methods of media text analysis. Cotter (2005, p. 416) argues that 'the discourse of news media encapsulates two key components: the news story, or spoken or written text, and the process involved in producing the texts.' She notes that the first dimension has been closely studied by many scholars, whilst the second one is often overlooked. This study of the Russian case of media coverage of environmental issues incorporates both of these dimensions. It looks at the complex production process of news, but at the same time, by means of content and discourse analysis, it studies the outcome of this process – written or spoken text produced by journalists under or in spite of the influence of the surrounding context. Content and discourse analysis can be interpreted and used very differently depending on the purpose of specific research; therefore, the use of these methods here will be outlined below.

Being a very popular method in studying media messages, content analysis allows the researcher to break the data into 'bits and pieces' (Pierce 2008) and gives a measure of 'quantifiability' to the project. Complexity of the content analysis depends on the particular case and purpose of the research and how researchers understand it. This study will use the definition given by Neuendorf (2002, p. 1): '[content analysis] defined as the systematic, objective, quantitative analysis of message characteristics', a characterisation which can be supplemented by Krippendorff's 1980 definition, which refers to content analysis as 'a research technique for making replicable and valid inferences from data to their context' (cited in Bertrand and Hughes 2005, p. 177).

Discourse analysis allows us to explore the uniqueness of the text produced by media (van Dijk 1991) where the meaning of words can be altered and understood completely differently depending on the context. So, discourse analysis will be used to look at the language as a whole, including the non-linguistic categories, such as who is presenting the news, who is the audience, what types of non-verbal communication are involved and how information is situated in the bigger context of social interactions. Discourse analysis has several interpretations,[2] hence there is quite a variety of schools and approaches to understanding and implementing discourse analysis, and even the definitions of discourse differ drastically. Van Dijk (2011, pp. 3–4) points out that discourse can be seen as 'social interaction, [. . .] as power and domination, [. . .] as communication, [. . .] as contextually situated, as social semiosis, as natural language use [or] [. . .] as a complex layered construct'.

Due to the specific nature of this research project (explained below), a methodology inspired by 'critical discourse analysis' (CDA) will be applied for this study, and in particular the work of two prominent linguists, Teun van Dijk and Norman Fairclough, will be drawn upon. Even though they differ in some of their views on discourse, in this case they could be referred to as representatives of one approach to discourse studies (Gillespie and Toynbee 2006). CDA considers discourse to be inseparable from its social context; however, as much as discourse is influenced and transformed by the surrounding interactions or environment, in turn it also possesses power and may change that context (Fairclough et al. 2011, p. 357). For this research, discourse will be seen as 'a form of social practices (economic, political, cultural and so on)' (Fairclough 2001, p. 122) which considers not only the nominative function of the language (defining the objects) but also the 'linguistic conceptualization of the world' (Fairclough et al. 2011, p. 358). Fairclough et al. (ibid) state that discourse

> may have major ideological effects [. . . It] can help produce and reproduce unequal power relations between (for instance) social classes, women and men, and ethnic groups, through the way it represents things and positions people.

In the case of climate change coverage in Russia, the following 'forms of social practices' will be considered – the state's position towards climate change policy, media dependency on the state, businesses' positions towards the problem and again media dependency on it, as well as the unclear messages produced by the scientific community, NGOs' struggle to lobby successfully on environmental issues, public reluctance to face the problem and the growing influence of international actors on Russia's climate change policy (see more in Chapter 3). So within this complex discourse created by various actors, journalists' choice of words acquires special functions – 'linguistic conceptualization'. For instance, if a state-owned newspaper after the Copenhagen Conference starts referring to climate change as a well-known fact rather than a lie created by Western scientists, then it can be argued that because of the political context, this choice of words is very likely to be connected with the state's policy, and journalists do not just show the change in their beliefs or knowledge (from unknown to known) but also the change of how the situation is perceived or 'conceptualised' by the main actors of the existing discourse.

Discourse analysis is also widely used by scholars studying media coverage of climate change (Boykoff 2008, Carvalho 2005, Carvalho and Burgess 2005, Doulton and Brown 2007, Fletcher 2009, Olausson 2009). When applied to media coverage of climate change issues in Russia, the discourse analysis closely resembles the main ideas of CDA. Russian journalists have to operate in the complex environment of the increasing governmental centralisation of media, diffused boundaries between big business and the state, the worsening environmental situation and alarming messages from NGOs and international communities. Carvalho and Burgess (2005, p. 1461) state that 'CDA attempts to understand the links between texts and social relations, distribution of power, and dominant values and ideas'. As can be seen, there are various social and power relations that coexist in post-Soviet society with the developing ideology of free market economics and the persisting ideology of the strong state. So, the analysis of Russian media will be conducted with consideration of this context. As Teun van Dijk argues, 'discourse is not produced without context and cannot be understood without taking the context into consideration' (cited in Fairclough et al. 2011, p. 372).

An important aspect of this research is that the 'green' media were purposely excluded from the investigation. Neither during the media analysis nor during the fieldwork were media that specialised solely in environmental issues approached. The rationale behind this decision came not from undermining the role and values of these types of media, but from the understanding that they by definition dedicate their time and efforts to raising awareness about problems such as climate change, hence their coverage will not be altered as much by state policy or any other external factors. At the same time, the 'green' media have quite specialised (often narrow) target audiences – people who are already concerned with environmental problems and the future of our planet. Whilst in the non-specialised popular media the climate change topic has to compete with many others, starting with economic news and ending with celebrity gossip, so the way journalists and editors prioritise the news and approach the problem provides rich material for analysis.

Elite interviews

The elite interviews became a substantial part of this research project. Considering the scarcity of literature on the subject, elite interviews allowed for an understanding of the interviewee's personal attitudes towards the problem as well as a reconstruction of events which were missing from the written information (Tansey 2007). Elite interviews suggest that a low number of respondents is substituted by the interviewees' high rank or their key position and deep knowledge of the subject. Overall, 31 interviews with journalists and specialists working in different types of media organs (newspapers, TV, radio and news agencies) were conducted, including representatives of environmental NGOs (Russian headquarters of major international environmental groups as well as NGOs working closely with government), policy makers involved in environmental control and climatologists who contribute to the development of Russian science on the subject but also provide consulting service to state officials. Half of the interviews were conducted in Moscow due to the significant political and economic influence of the Russian capital. However, considering Russia's vast geography and the substantial differences between the European part of the country – mainly Moscow and St. Petersburg – interviews were also conducted with experts based in Barnaul, Novosibirsk, Krasnoyarsk, Petrozavodsk, and Kemerovo. The interviews were carried out in person during fieldwork trips to Russia in July and August 2011, as well as by telephone, Skype, and emails throughout 2011–2012. The average length of the interviews was one hour.

A number of researchers outlined various mistakes or obstacles which could be encountered during the process of elite interviewing. Among them were difficulties with getting interviews in the sense of identifying the key figures or getting access to high-profile policy makers (Goldstein 2002); defining the purpose and structure of the interview (Aberbach and Rockman 2002, Leech 2002); choosing and constructing the questions (Berry 2002); and being attentive to ethical aspects of interviews (Woliver 2002). Conducting interviews in Russia also involves other specific problems which to a certain degree were reflected by Werning Rivera et al. (2002). These include difficulties in gaining access to key people and arranging interviews (the absence of secretaries or personal assistants for politicians or inefficient use of emails are important challenges), less experience with the interview process (in comparison with Western countries) and unfamiliarity with academic research (it is difficult to explain the purpose of your inquiry and to get an adequate response).

All of these problems were experienced and resolved during this research project. Interviews were arranged in advance through emails or telephone calls. The nature and purpose of academic research was explained beforehand, with full disclosure of how the material would be used. All interviewees were asked about their preference for anonymity (in the majority of cases, the identities of the interviewees will be hidden throughout the text of this book), and they were informed that the transcripts of interviews and any published material would be provided upon request.

In addition to the above-mentioned challenges with conducting interviews, researching climate change media coverage presents another specific obstacle: there is an extremely limited number of people with sufficient knowledge on the subject. During the data collection stage, only four journalists were identified who regularly worked on the problem of climate change in Russia. In other cases, journalists were either writing generally on environmental topics or randomly covering climate change without any specialised knowledge. In the case of policy makers, the situation was even more pessimistic, as will be discussed in detail in Chapter 3 – in Russia there is a very vague understanding of which institution is in charge of climate policy, hence once again it makes it complicated to find the 'right' person with sufficient knowledge of the problem (unfortunately, high-ranking politicians involved in this process were not reached). An extremely valuable source of information were NGO representatives. Due to their diversity and deep understanding of the problem, they not only provided the vision of the problem as activists, but also themselves acted as journalists, news sources, and scientists (many have advanced academic degrees), and some of them actually contribute to developing Russia's climate policy or take part in international negotiations as members of Russia's official delegation.

Additionally, official documents – such as Russian federal and regional laws, presidential decrees, expressions of state doctrines, reports prepared by various state agencies and so on – became important sources of data for this research. Chapters 3 and 5 rely on the content analysis of official presidential speeches made publicly available on the Russian president's website.

Research focus and terminology

Conducting research on any dimension of climate change issues tends to suggest a clear understanding of the problem itself. Being a social scientist rather than a natural scientist implies that the researcher has to accept the science of the problem for granted and rely on secondary data provided by the international community of climatologists. A brief overview of how this scientific problem is understood in this research project is hereby presented.

The Earth's climate is regulated through a balance of energy received from the sun and energy emitted back to space. Atmospheric gases and clouds are responsible for trapping the energy which is reflected off the Earth's surface, leading to the 'greenhouse effect' (The Royal Society 2010). The research shows that the largest contributors to the 'greenhouse effect' are water vapour and CO_2. Climate change, provoked by the modifications in this balance, is the manifestation of the planetary system trying to 'adjust' and regulate the Earth's temperature. According to the Royal Society (ibid), changes in climate are evident throughout Earth's history due to various natural phenomena. The impact of human activity on these changes is, however, quite a recent cause of climate change that has upset the Earth's ability to naturally regulate the climate system.

This research project is concerned with the media coverage of *anthropogenic* climate change. The international scientific community has reached a consensus that anthropogenic emissions of GHGs from the burning of fossil fuels, deforestation, and industrial and agricultural activities are largely responsible for the increase in the average temperature of the Earth by 0.85°C over the past century. Unlike natural phenomena such as volcanic eruptions and sustained variations in the energy emitted by the sun (The Royal Society 2010), anthropogenic climate change can be mitigated by people through reducing GHG emissions. Considering the world's dependence on fossil fuel energy and the global nature of climate change's consequences, this process involves a clash of interests where states are supposed to become the main actors or negotiators in the climate change mitigation process.

Another clarification which needs to be addressed with regard to the case study of this research is the confusion between the terms 'climate change' and 'global warming'. Most of the time in mass media these terms are used interchangeably as synonyms. Strictly speaking, 'global warming' refers to the overall trend of rising temperatures, whilst 'climate change' is a more general term that includes an increased frequency of extremely cold or hot seasons, increases or decreases in the amounts of precipitation, increases in anomalous weather events (hurricanes, droughts, snow storms and so on). Even though these climatic changes still happen within the context of the overall rise in temperature, in terms of media coverage, the use of the term 'climate change' signals certain trends in understanding the problem. Carvalho in 2006 (see in Good 2008) stated that until the end of the 1980s the term 'greenhouse effect' prevailed in the public discourse, but by the early 1990s it was replaced by 'global warming'. Good (2008) points out that now the dominance of the term 'climate change' can be seen even through the name of the Intergovernmental Panel on Climate Change (IPCC) rather than the Intergovernmental Panel on Global Warming (see more on this issue in Linder 2006, Schuldt et al. 2011, Whitmarsh 2009). Russian climatologist Nataliya Kharlamova (presentation, Chemal,[3] 13 August 2011), pointed out that just five to seven years ago people were talking about 'global warming'; however, when Russia recently experienced severely cold winters, the term 'global warming' was ridiculed and the temporary drop in temperature was used as evidence of its falsehood. Then, according to Kharlamova (ibid), for scientists it was a reason to start promoting the term 'climate change', which would cover a broader range of natural abnormalities. In this research project, during data collection both terms were used (climate change and global warming [*izmenenie klimata, global'noe poteplenie*]) due to the scarcity of information on the topic, this approach being aimed to broaden the results of the scientific inquiry.

The book's structure

The outlined problems addressed by the research are depicted in the following chapters:

1   **Mass media and climate change: Its role, challenges and trends** – this chapter reviews existing studies on media and climate change. It explores the dilemma of the studies of media reporting on climate change which from

one side involves complex scientific communication, and from another side, considering the impact and complexity of climate change, involves analysis of various factors influencing this coverage. Even though the analysis demonstrates that the considerations of micro-processes (for example, widely discussed journalistic norms and practices) were studied in greater detail, recently scholars more often turn their attention to the macro-factors influencing media coverage of climate change (such as politics and the economy).

2   **The political economy of Russian mass media: State and market** – this chapter presents an analysis of media in Russia and explores the major modifications they went through over the last decades. The chapter looks at various components of the political economy of Russian media: ownership structure, advertisement, sources of information, flak (censorship) and dominant ideology of the state. By 'deconstructing' the media system into these components, the chapter unfolds the possible barriers which the climate change topic encounters in Russia.

3   **Russian climate change policy: Towards 'climate pragmatism'** – this chapter looks at the development of Russian climate policy from being mostly disinterested in the subject matter to becoming more concerned with the problem (which coincided with Medvedev's presidency). Content analysis of Medvedev's official speeches shows that Russia's leaders now see climate policy as being in the best interest of the state and leading economic actors. Indeed, we can already observe this move towards the policy of *climate pragmatism*, where carbon emissions are cut through the modernisation of the economy and improvement in energy efficiency.

4   **Russian newspapers and climate change** – this chapter offers an analysis of Russian press coverage of climate change. The data are collected from five national newspapers: *Izvestiya* (right-wing newspaper), *Kommersant* (liberal), *Rossiyskaya gazeta* (state owned), *Komsomol'skaya pravda* (tabloid) and *Sovetskaya Rossiya* (communist). The aim of this chapter is to study the dynamics of media coverage by looking at how the amount and the character of climate change news has changed over time depending on certain conditions (modifications in state policy, global conferences on climate change, acceptance of international documents and so on). Coverage in these newspapers will be studied by focusing on three events: the Kyoto Conference, the Copenhagen Conference (plus acceptance of the Climate Doctrine) and the heat wave of 2010 in Russia.

5   **Mediating climate change in Russia: Passing through the barriers** – this chapter explores the key role of the state and economic elites in determining climate change coverage in Russia and discusses the lesser, but still important, role of micro-factors (such as the specifications of the topic, the influence of journalistic professional norms and the role of experts). Furthermore, this chapter highlights the fact that the research findings show not only a biased media policy towards the problem, but also (at times) the omission of the issue from the media discourse. The chapter finishes with a discussion of how media coverage of climate change in Russia can be improved.

Notes

1 According to Louw (2010, p. 31), 'the notion of the Fourth Estate media' originates in Locke's 'free-flow of information principle' (which he discussed in *Second Treatise of Government* [1966]). From this point of view, media become an essential element of liberal democracy which allows the public to control the other branches of power and expose them if they deviate from the principles of the democratic state.
2 Schiffrin et al. (2005, p. 1) note that generally for linguists, 'discourse' means anything 'beyond the sentence', whilst for other researchers, *discourse analysis* is an analysis of 'language use'. While both these understandings of discourse involve language, critical theorists go beyond these definitions and look not only at the linguistic but also at the non-linguistic characteristics of discourse (the social context in which language has to operate).
3 Referencing of personal communication includes geographical location of where the communication took place rather than where the interviewees are based.

Bibliography

Aberbach, J. and Rockman, B. (2002) 'Conducting and Coding Elite Interviews', *PSOnline*, www.apsanet.org, date accessed 29/06/2011.

Berry, J. (2002) 'Validity and Reliability Issues in Elite Interviewing', *PSOnline*, www.apsanet.org, date accessed 29/06/2011.

Bertrand, I. and Hughes, P. (2005) *Media Research Methods. Audiences, Institutions, Texts*, Basingstoke: Palgrave Macmillan.

Boykoff, M. (2008) 'The Cultural Politics of Climate Change Discourse in UK Tabloids', *Political Geography*, 27: 549–569.

Carvalho, A. (2005) 'Representing the Politics of the Greenhouse Effect: Discursive Strategies in the British Media', *Critical Discourse Studies*, 2/1: 1–29.

Carvalho, A. (2008) 'Media(ted) Discourse and Society', *Journalism Studies*, 9/2: 161–177.

Carvalho, A. and Burgess, J. (2005) 'Cultural Circuits of Climate Change in UK Broadsheet Newspapers, 1985–2003', *Risk Analysis*, 25/6: 1457–1469.

Charap, S. (2010) 'Russia's Lacklustre Record on Climate Change', *Russian Analytical Digest*, 79: 11–15.

Cotter, C. (2005) 'Discourse and Media'. In: D. Schiffrin, D. Tannen and H. Hamilton (eds.) *The Handbook of Discourse Analysis*, Oxford: Blackwell Publishing: 416–436.

Doulton, H. and Brown, K. (2007) 'Ten Years to Prevent Catastrophe? Discourse of Climate Change and International Development in the UK Press', *Tyndall Centre Working Paper* 111: 1–58.

Doyle, A. (2009) 'Russia CO_2 Emissions Up in 2007, Lag GDP Growth' (16 April), www.reuters.com/article/environmentNews/idUSTRE53F1HR20090416, date accessed 27/04/2009.

EIA [US Energy Information Administration] (2013) *Country's Analysis Briefs: Russia* (last updated 26 November), www.eia.gov/countries/cab.cfm?fips=RS, date accessed 3/06/2015.

Fairclough, N. (2001) 'Critical Discourse Analysis as a Method in Social Scientific Research'. In: R. Wodak and M. Meyer (eds.) *Methods of Critical Discourse Analysis*, London: Sage.

Fairclough, N., Mulderrig, J. and Wodak, R. (2011) 'Critical Discourse Analysis'. In: T. van Dijk (ed.) *Discourse Studies: A Multidisciplinary Introduction – Volume One*, London: Sage.

Fletcher, A. L. (2009) 'Clearing the Air: The Contribution of Frame Analysis to Understanding Climate Policy in the United States', *Environmental Politics*, 18/5: 800–816.

Giddens, A. (2010) 'Russia Can Ill-Afford Climate Cavalierism', *Policy Network* (31 August), www.policy-network.net/pno_detail.aspx?ID=3884&title=Russia-can-ill-afford-climate-cavalierism, date accessed 13/12/2010.

Gillespie, M. and Toynbee, J. (2006) *Analysing Media Texts*, Maidenhead: Open University Press.

Goldstein, K. (2002) 'Getting in the Door: Sampling and Completing Elite Interviews', *PSOnline*, www.apsanet.org, date accessed 29/06/2011.

Good, J. (2008) 'The Framing of Climate Change in Canadian, American, and International Newspapers: A Media Propaganda Model Analysis', *Canadian Journal of Communication*, 33: 233–255.

Henry, L. and Douhovnikoff, V. (2008) 'Environmental Issues in Russia', *Annual Review of Environment and Resources*, 33: 437–460.

Leech, B. (2002) 'Asking Questions: Techniques for Semistructured Interviews', *PSOnline*, www.apsanet.org, date accessed 29/06/2011.

Linder, S. (2006) 'Cashing-in on Risk Claims: On the For-Profit Inversion of Signifiers for 'Global Warming', *Social Semiotics*, 16/1: 103–132.

Louw, E. (2010) *The Media and Political Process*, 2nd ed., London: Sage.

Muenchmeyer, T. (2008) 'Less Money for Fur Coats – Ignorance and Arrogance in Russian Climate Policy', *Osteuropa*, 58: 217–236.

Neuendorf, K. (2002) *The Content Analysis Guidebook*, London: Sage.

Oates, S. (2007) 'The Neo-Soviet Model of the Media', *Europe-Asia Studies*, 59/8: 1279–1297.

Olausson, U. (2009) 'Global Warming – Global Responsibility? Media Frames of Collective Action and Scientific Certainty', *Public Understanding of Science*, 18/4: 421–436.

Pearce, F. (2003) 'Global Warming Will Hurt Russia' (3 October), www.newscientist.com/article/dn4232-global-warming-will-hurt-russia.html, date accessed 5/11/2012.

Perelet, R., Pegov, S. and Yulkin, M. (2007) 'Climate Change. Russia Country Paper', *Human Development Report 2007/2008*. http://hdr.undp.org/sites/default/files/perelet_renat_pegov_yulkin.pdf, date accessed 20/03/2010.

Pierce, R. (2008) *Research Methods in Politics: A Practical Guide*, London: Sage.

President of Russia website (2010) 'Zasedanie Soveta Bezopasnosti po Voprosam Izmeneniya Klimata', www.kremlin.ru/news/7125, date accessed 1/03/2011.

Schuldt, J., Konrath, S. and Schwarz, N. (2011) ' "Global Warming" or "Climate Change"? Whether the Planet Is Warming Depends on Question Wording', *Public Opinion Quarterly*, doi: 10.1093/poq/nfq073, pp. 1–10.

Schiffrin, D., Tannen, D. and Hamilton, H. (eds.) (2005) *The Handbook of Discourse Analysis*. Oxford: Blackwell Publishing.

Tansey, O. (2007) 'Process Tracing and Elite Interviewing: A Case for Non-probability Sampling', *PS: Political Science and Politics*, 40/4: 765–772.

Tekin, A. and Williams, P. (2009) 'EU-Russian Relations and Turkey's Role as an Energy Corridor', *Europe-Asia Studies*, 61/2: 339–352.

The Royal Society (2010) *Climate Change: A Summary of the Science*, London: The Royal Society.

Tynkkynen, N. (2010) 'A Great Ecological Power in Global Climate Policy? Framing Climate Change as a Policy Problem in Russian Public Discussion', *Environmental Politics*, 19: 179–195.

UNESCO (2009) *A Commitment to Act Now. Broadcast Media and Climate Change*, Paris: UNESCO.

van Dijk, T. (1991) 'The Interdisciplinary Study of News as Discourse'. In: K. B. Jensen and N. Jankowski (eds.) *A Handbook of Qualitative Methodologies for Mass Communication Research*, London: Routledge: 108–120.

van Dijk, T. (2011) *Discourse Studies: A Multidisciplinary Introduction – Volume One*, London: Sage.

Werning Rivera, S., Kozyreva, P. and Sarovskii, E. (2002) 'Interviewing Political Elites: Lessons from Russia', *PSOnline*, www.apsanet.org, date accessed 29/06/2011.

Whitmarsh, L. (2009) 'What's in a Name? Commonalities and Differences in Public Understanding of "Climate Change" and "Global Warming"', *Public Understanding of Science*, 18: 401–420.

Wilson Rowe, E. (2009) 'Who Is to Blame? Agency, Causality, Responsibility and the Role of Experts in Russian Framings of Global Climate Change', *Europe-Asia Studies*, 61/4: 593–619.

Woliver, L. (2002) 'Ethical Dilemmas in Personal Interviewing', *PSOnline*, www.apsanet. org, date accessed 29/06/2011.

Yagodin, D. (2010) 'Russia: Listening to the Wind – Clientelism and Climate Change'. In: E. Eide, R. Kunelius and V. Kumpu (eds.) *Global Climate – Local Journalisms. A Transnational Study of How Media Make Sense of Climate Summits*, Bochum: Projektverlag: 275–290.

Zassoursky, I. (2004) *Media and Power in Post-Soviet Russia*, New York: M. E. Sharpe.

1 Mass media and climate change

Its role, challenges and trends

In the last few decades we can observe how the studies of media coverage of climate change have generated tremendous interest within the scholarly community. When climate change emerged as a key research issue, most research projects focussed on the West (e.g., Dirikx and Gelders 2009, Gavin 2009, Hulme 2009, Lockwood 2009, Lyytimaki 2011). However, now studies are conducted all over the world, including in the largest economies and polluters such as China, India and Canada (Davidsen and Graham 2014, Shanahan 2009, Wu 2009, Xu 2010), as well as in the most vulnerable countries already facing the consequences of our changing climate, such as Bangladesh, Pakistan and Nigeria (Akpan et al. 2012, Rhaman 2010, Shanahan 2009). Moreover, scholars have moved forward and proposed conducting comparative studies among various countries (Grundmann and Scott 2014, Schafer et al. 2014, Schmidt et al. 2013) and within the same country (Liu et al. 2008).

The importance of the media in communicating environmental risks has been stressed by Ulrich Beck (1994, p. 23) in his influential monograph on risk society in which he states: '[risks[1]] can be changed, magnified, dramatized or minimized within knowledge, and to that extent they are particularly open to social definition and construction. Hence the mass media and the scientific and legal professions in charge of defining risks become key social and political positions.' Beck (ibid, p. 197) stresses that 'expensive and extensive scientific investigations are often not really noticed in the agency that ordered them until television or a mass-circulation newspaper reports about them.' The high level of interest in the media coverage of climate change can be explained by the crucial role the media play in translating the abstract threats of climate change as reported by science into the language of the general public (Antilla 2005, Boykoff and Boykoff 2007, Carvalho 2007, Carvalho and Burgess 2005), in forming people's opinions (Boyce and Lewis 2009, Sampei and Aoyagi-Usui 2009, Taddicken 2013), in shaping perceptions and reactions to the danger posed by climate change (Boykoff 2012, Lockwood 2009, Pasquare and Oppizzi 2012), in serving as middlemen between the people, science, business and policy makers (Butler and Pidgeon 2009) and in ascribing responsibility for the problem (Olausson 2009). Analysis of media coverage of the 'risks' associated with climate change helps us to understand why some narratives become salient and some remain so (Boykoff 2008a) and what factors or actors shape the created discourse.

In some political regimes the mass media are powerful enough to influence how the leading elites and population alike understand and respond to climate change (Boykoff 2008a). However, often mass media do not fulfil these beneficial functions, but on the contrary create barriers or obstacles by distorting the information or approaching the topic from a questionable angle or by simply ignoring or undermining discussion of the climate change problem (Antilla 2005, Boykoff and Boykoff 2004, Carvalho 2007). This chapter critically assesses existing studies on media coverage of climate change, and is divided into three sections: first, the specification of the climate change topic; second, how journalists' professional values shape coverage; and finally, the role played by politics in climate change coverage.

Climate change as a topic

Climate change is indeed one of the biggest challenges of our time and the media are capable of playing a crucial role in popularising the danger of climate change among the wider public. Before we embark on an analysis of how the media deal with this task and what social, economic or political barriers they encounter, it is crucial to realise that climate change is an unusual topic and on its own it has the potential to become a barrier for journalists. As with other environmental topics, climate change is an 'unobtrusive issue' (this term was introduced by Atwater and colleagues in 1985 (cited in Shanahan and Good 2000). Whilst 'obtrusive issues' such as economic recessions are clearly evident for people and directly affect their lives, climate change is not that apparent and straightforward and demands that journalists do 'an extremely difficult job' connecting 'global warming, weather extremes, flooding and human activity' (Gavin et al. 2011, p. 433). In this regard, it is useful to refer to Schumpeter's (1943) discussion of the 'classical doctrine of democracy', in which he reaches the following conclusion about foreign news: 'these things seem so far off; they are not at all like a business proposition; dangers may not materialize at all and if they should they may not prove so very serious, one feels oneself to be moving in a fictitious world' (Schumpeter 1943, p. 261). Further on he continues:

> The reduced sense of responsibility and the absence of effective volition in turn explain the ordinary citizen's ignorance and lack of judgement in matters of domestic and foreign policy which are if anything more shocking in the case of educated people and of people who are successfully active in non-political walks of life than it is with uneducated people in humble situations. Information is plentiful and readily available. But this does not seem to make any difference.
>
> (Schumpeter 1943, p. 261)

Here, if 'foreign news' is replaced by 'climate change', then Schumpeter's arguments remain relevant without any modifications. For an ordinary person, climate change is an abstract idea, and even though it could be stated that there is a very limited number of people who are unaware of climate change or global warming, the rudimentary understanding of the problem prevails, as well as the detached perception of its effects.

It often gets forgotten that it is due to the collaborative work of scientists around the world that climate change was discovered. As Dorothy Nelkin (1995, p. 2) states, 'for most people, the reality of science is what they read in the press'. Climate change, like any other scientific topic, represents 'an encoded form of knowledge that requires translation in order to be understood' (Boykoff and Boykoff 2004, p. 126, citing Ungard 2000, p. 308; also see Dirikx and Gelders 2009), which involves joint work between scientists and journalists. For journalists to be able to provide proper coverage of the topic, they need to have at least some understanding of the problem, otherwise 'public confusion is exacerbated by reporters who misunderstand the basic scientific principles of climate change' (Antilla 2005, p. 350, citing Wilson 2000a). Ideally, journalists who specialise in climate change need to have some training or initial background knowledge which will allow them to have a grasp of the development of the science of climate change as well as its politics (Shanahan 2007).[2] Scientists are also responsible for communicating their findings in an adequate way by adding 'authority and legitimacy to environmental reporting' (Taylor and Nathan 2002, p. 330). Carvalho (2007, p. 228) argues that in the 1980s 'scientists were the uncontested central actors and exclusive definers of climate change' (she refers to original studies conducted by Boykoff and Boykoff 2004, McComas and Shanahan 1999, and Trumbo 1996). As will be discussed below, this situation changed quite quickly in the early 1990s when scientists became more reticent in their communication with journalists and due to the growing complexity of the problem, scientists were replaced by politicians or economists (as information sources).

As Smith (2005) confirms through his research, journalists have often directed accusations at scientists who do not completely fulfil their duty of assisting in communicating climate change threats. For instance, one of the common misunderstandings between scientists and journalists is the probability of climate change actually taking place (Keeling 2009). Journalists need a clear answer as to whether climate change is happening or not and what the human impact on it is. However, in scientific discourse, universal agreement is practically impossible, and that is why statements in IPCC reports arguing that anthropogenic climate change is happening tend to suggest that it is very likely (say, a more than 90 percent chance). As James Painter (2013, p. vii) puts it, 'many people fail to recognise the distinction between "school science", which is a source of solid facts [. . .], and "research science" where uncertainty is engrained'. This distinction is quite understandable for scientists and accepted as part of their work routine but needs 'translation' for journalists.

Another very common problem is that often journalists and the broader public do not differentiate between 'climate' and 'weather'. As Bostrom and Lashof (2007) argue, this mistaken substitution of the concepts leads to different behavioural and policy outcomes. 'Weather' is seen by people as a natural phenomenon which cannot be influenced and which is taken for granted, hence there is no responsibility for humans. Furthermore, when people see climate change as a change in weather, then after one cold month or a 'normal' season in terms of temperature, we can see in the press such claims as 'after this [cold] winter [the] number of people doubting global warming is significantly increasing' (Bersenev 2009), whilst media coverage supporting climate change's existence is highly criticised for 'fooling' people – 'the art of propaganda has reached that state, when even extremely cold weather is explained by global warming' (Latynina 2010).[3]

Writing about climate change, journalists have to deal with a strong (or one could rather say 'loud') lobby of climate sceptics (see Gavin and Marshall 2011). Arguably because the topic is so controversial, the media become even more important in the way climate change is presented and transmitted to the audience (Carvalho 2007). As will be discussed below at length, journalists often struggle to identify how fair the sceptics' attacks are and the strength of the disagreement inside the scientific community about anthropogenic climate change. Arguably it happens again due to the lack of adequate qualifications or training, as well as the nature and routine of journalists' work (Stocking and Holstein 2009). When journalists are not able on a daily basis to conduct in-depth investigative analysis on who is right or wrong and which science is funded by whom, eventually journalists themselves become suspicious of climate change science or twist the topic in a way that fits their interests.

A related problem which also shapes journalists' coverage of the highly scientific topic of climate change is that journalists tend to employ 'heuristics' that are different than those employed by scientists. As Kahneman and colleagues explained, 'thinking about risks, people rely on certain heuristics, or rules of thumb, which serve to simplify their inquiry' (Sunstein 2006, p. 198, citing Kahneman et al. 1982). Dunwoody and Griffin (2002) state that it is to a large extent due to this hectic work routine of journalists (where they cannot afford to spend too much time on one issue) that they are fast to use 'judgemental shortcuts' in the process of making up their mind on complex problems:

> In a world of rapidly recurring deadlines, journalists cannot afford to engage in systematic information processing. Instead, the occupation rewards those who can make quick decisions about 'what's news' and decide rapidly how to cobble together a story. Extremely fast decisions are, perforce, heuristic ones. Thus, journalism is unapologetically a world of heuristic decision making.
>
> (ibid, p. 180)

For instance, Dunwoody and Griffin (2002) provide several examples of journalists' heuristic in defining 'news value'. Firstly, they state that for journalists, 'size matters', meaning that the larger the impact of the event, the more likely it will become of interest to journalists. Secondly, 'the closer, the better': if something happens in direct proximity to journalists' audiences, it will be prioritised over events taking place across the globe. Lastly, once something becomes news, it is very likely that it will remain news for some time. Other examples of journalists' heuristics involve the prioritisation of events over the process, negative information over positive information or personal illustration of information. In the case of climate change, the last point would mean that journalists prefer to build their news stories around an interview with a person who witnessed a polar bear dying or noticed how energy bills drastically decreased/increased rather than 'utilising systematic or consensus data' (ibid, p. 187) provided by a group of climatologists. The authors also argue that these heuristics are not unique to journalists, but are 'used by most individuals to negotiate daily life[. . . .] Journalists' practice reinforces reporters and editors for using heuristics that are integral to problem solving for all of us' (ibid, p. 178). The specifications of journalists' professional habits or norms will be discussed further on with regard to climate change reporting.

Journalists are not alone in their desire to use scientific knowledge in a way that is convenient for them; as Carvalho and Burgess (2005) argue, politicians also look for ready answers from scientists for further interpretation of the scientific knowledge in order to fit it within their political agenda. For example, former British prime minister Margaret Thatcher used the scientific uncertainty on the subject as a justification for the government's inaction on the climate change matter and appealed to the public to wait in order to not make a mistake (Carvalho 2005). Hansson (2004, p. 357) characterises this decision-making pattern as 'the delay fallacy' – 'if we wait we will know more about X: no decision about X should be made now'. However, in the case of public decision making on risks (such as climate change), this reaction leads to an obvious problem – while waiting, the risk gets worse. Hansson himself refers to this as 'one of the most dangerous fallacies of risk' (ibid), since resolving the scientific uncertainty on the issues would imply reaching the ultimate position of possessing all the knowledge on the subject, which becomes near to an impossible situation (as was noted beforehand, scientific certainty is extremely difficult to achieve), then the 'delay fallacy' argument 'can almost always be used to prevent risk-reduction actions' (Hansson 2004, p. 357).

Scientific knowledge is also questioned by businesses and activists involved in the climate change regulation processes, and as a consequence 'science has become more exposed to criticism, contestation and deconstruction' (Carvalho 2007, p. 224). All of these cases of the misinterpretation, twisting or scrutiny of scientific information leads to the problem of scientists becoming more and more aware of their vulnerability (Keeling 2009) and being cautious in what they say and how they say it.[4]

It is now apparent that climate change is not only a scientific or even environmental topic, but instead the public comprehends it as a more complex phenomenon which coexists with other socioeconomic and political problems (Zehr 2009). Nielsen and Kjaergaard (2011, p. 26) share similar ideas and state that 'climate change [is] making the transition from a scientific hypothesis to an established fact in public debates, the issue [becoming] political, financial, and ethical as much as scientific', and journalists' roles in interpretation or the framing of these issues becomes to some extent an economic or political tool. Further on, journalists' professional norms will be discussed with regard to the way they shape media coverage of climate change.

Journalists' views of climate change

Regardless of the specification of climate change as a topic (whether it is a scientific, political or economic issue), some argue that in order for journalists to start writing about climate news, they need to see 'news value' in it, or as Carvalho (2007, p. 224) argues, 'novelty, controversy, geographic proximity and relevance for the reader, for example, are important determinants in the selection of science news'. Maxwell Boykoff and Jules Boykoff have made a significant contribution to the development of the study of media coverage of climate change. One of their most widely held arguments revolves around the idea that media coverage of climate change is determined by the journalists' professional habits or norms (for example, see Boykoff 2007, 2008b; Boykoff and Boykoff 2004, 2007), which become a natural barrier in the way of transforming scientific description of the process in its professional manner – with an unavoidable amount of uncertainty – into comprehensible information for the mass public.

Journalistic norms include first-order norms (personalisation, dramatisation and novelty) and second-order norms (authority, order and balance) (Boykoff and Boykoff 2004). First-order norms are superior in the sense that if the conditions are met (the news is dramatic and novel), news articles are most likely to be published, and then the second-order norms would in their turn influence *the way* the articles are published (Boykoff and Boykoff 2007). These norms not only serve as the barriers to climate information getting on the pages of newspapers, but also allow journalists to twist information or create 'informational bias', which in turn might give policy makers the opportunity to postpone their reaction to the problem. These norms will be briefly discussed below.

The norm of 'personalization' characterises the journalists' skills to portray abstract global political or economic processes through stories about particular individuals (Boykoff and Boykoff 2007), which helps people to visualise the problem and make connections between the causes and consequences. Shanahan (2007, p. 2) suggests that people will be more interested in the climate change topic if it is 'framed to suit diverse audiences'. Animal welfarists might be touched by disturbing images of a drowning or starving polar bear, while other people might be interested in the national security or economic side of the problem (Howard-Williams 2009 citing Lowe 2006). Similarly, people are more likely to react to climate change problems if they are 'defined in ways that emphasise local and regional impacts' (Pralle 2009, p. 791). On the other hand, if people see climate change as something distant from their everyday lives, they are less likely to react to it.

A similar concept of how journalists convey abstract information on climate change is presented by Hoijer (2010), who suggests that journalists use the communicative mechanism of 'anchoring and objectifying' in their coverage of climate change. Through anchoring, journalists bring unknown concepts into the known context – for instance, a comparison of climate change to mad cow disease. Objectification is extremely similar to personalisation, but it makes an abstract concept more concrete through relating it to not only individuals, but also animals (once again, polar bears) and even the observable process of the melting Arctic ice, basically anything which makes the problem more visual. Perhaps this norm could be considered as a positive strategy used by journalists, allowing them to make an issue more relatable to the audience and help to avoid the problem of 'distanciation', defined by McManus (2000) as the situation where climate change-related issues and events that are described by the media do not make any links to the everyday lives of the public. From another side, the danger of overly relying on the 'personalisation' norm is that the topic is very likely to disappear from the public arena when, let's say, Obama, Cameron or Britney Spears have nothing to say about it or the judgement of the problem becomes identical with the people's perception of the public figure.

The norm of 'dramatisation' allows journalists to bring a remarkable brightness to their work and catch the attention of their audience. Arguably, climate change is already an extremely dramatic event with severe (mostly unavoidable) consequences for the whole world; however, the natural prolonged drama of climate change is often not enough for journalists, and they tend to exaggerate the reality and focus on the sensationalist nature of the problem. James Painter in 2007 (cited in Shanahan 2007) compared the TV coverage of IPCC reports in China, Brazil, Russia, Mexico and South Africa and concluded that the more negative first IPCC reports on the consequences of climate change got much greater coverage than the second reports on the mitigation of climate change impact. However, journalists often go to even greater extremes and publish articles under such titles as 'Global Warming Will Bring Black Plague Back to Europe' (Busygin 2005) and 'Humanity Will Not Survive 21st Century' (Karavaev 2005), which openly claim that at this point there is nothing people can do, the end of the world is coming and we cannot even postpone it. Partly, the problem comes again from journalists neglecting the uncertainty of scientists' prognoses. It is possible in some scenarios that climate change could kill every human on Earth. But this modality of 'possibility' rather than certainty does not make a catchy headline. This kind of coverage causes a number of problems, such as leaving the audience with two possible options: we should not do anything because it will not change anything or we should not do anything because it is too absurd to be true. In both cases, the coverage does not provoke any constructive behaviour from its readers. From another side, it makes the climate change topic more vulnerable to the sceptics' attacks.

There is another way for journalists to talk about climate change in an evidently dramatic, sensational way without distortion or misrepresentation of it – to connect the coverage with natural disaster events which sometimes directly or indirectly relate to climate change and make it extremely visible and relevant to the media audience. For instance, Spence et al. (2011, p. 48) demonstrate that in the UK, people who have experienced severe weather events (such as flooding) are more tuned in to environmental messages. However, this approach triggers more problems than solutions, since from the scientific point of view the correlation is quite questionable. For example, weather abnormalities per se do not prove the existence of climate change – at the same time, the progression of climate change involves the possibility of an increased frequency of extreme weather events (see more in

Devine 2012). The research conducted by Shanahan and Good (2000) on the connection between the amount of media coverage of climate change and variations in temperature has demonstrated that there are some relationships between these two variables, but they conclude that temperature is a much weaker factor than the science or politics surrounding climate change issues. Of course, one might argue that temperature variation is not as definite as natural disasters.

The last first-order norm, 'novelty', asks journalists to always report brand-new information, whereas the continuing problem of climate change might stop being newsworthy. Anthony Downs (1972, p. 38) argues that since it is difficult to keep the public's attention on one problem for a long period of time, information tends to go through an 'issue-attention cycle'. According to this theory, developing a story on the news does not reflect real developments but the development of *people's attention to the story*. The cycle contains five stages. The first is the 'pre-problem stage': only small groups of people are aware, such as experts or involved groups. The second stage is 'alarm', when the public suddenly gets a chance to learn about the problem and is quite hopeful about solving it without major losses. In the third stage, people start to see the problem in a realistic way and realise that a solution might demand great efforts and resources. When more and more people start to understand the actual sacrifices and work that they need to do in order to overcome the problem, the issue-attention cycle reaches the fourth stage. At this point the public gets scared, unmotivated and simply bored. The final stage is when the problem is no longer a centre of attention, but at the same time it does not disappear from people's lives. Periodically it attracts attention again, mostly because during the stage of biggest attention, some organisations are created or documents signed to solve the problem and the activity of these organisations or the conditions of these agreements periodically create news.

In the case of media coverage of climate change, in order for the topic to achieve the second stage of 'alarm', something extraordinary has to happen in front of the public's eyes: a natural disaster like flooding or forest fires. However, the nature of the problem of climate change is such that it happens constantly and often does not reveal itself through weather extremes but rather through gradual changes such as temperature increases. Boykoff and Boykoff (2007) argue that their content analysis does not prove the adequacy of Downs's model for climate change issues. The model does not take into consideration the role of the mass media, as well as other factors which influence the coverage of climate change, such as political and economic processes, and even, as Zehr (2009, p. 81) argues, 'the controversy and uncertainty that was constructed around it.' Downs suggests that consideration should only be given to the characteristics of the environmental topic itself, which will naturally go through this issue-attention cycle, whilst Boykoff and Boykoff (2007, p. 1195) state that climate change over time becomes a more and more serious problem, hence they conclude that 'the persistence of environmental problems on the social docket is affected more by the way these problems are constructed in the news media than by a natural history framework'.

The second-order journalistic norm of 'authority-order' describes the journalists' desire to refer to the powerful players as sources of trustful information:

'authority figures – government officials, business leaders, and others – who reassure the public that order, safety, and security will soon be restored' (Boykoff and Boykoff 2007, p. 1193, citing Bennett 2002). Howard-Williams (2009) points out that for such abstract and complicated issues, the choice of sources plays a particularly important role. This norm creates a problem when the sources of information for the scientific problems of climate change become not climatologists themselves but policy makers, state leaders and heads of major energy corporations,[5] suggesting that the more powerful source is the better positioned to offer explanations of climate issues, which in many cases is not true. The problem of politicisation of the climate change topic will be discussed further on in this chapter.

The journalistic norm of balance deserves a separate discussion, since arguably after the climate change topic reaches the news pages, *balance* becomes one of the most 'damaging' journalistic norms with regard to the media coverage of climate change.

Balancing climate

One of the most basic professional principles that journalists around the world follow in their everyday practice is that of objectivity and a balance of facts, which is presented in their works by different and sometimes contradictory news sources. Firstly, a professional reporter needs to detach himself from the problem (Ward 2008), so that his personal feelings and ideas do not interfere. Secondly, journalists are expected to 'present the views of legitimate spokespersons of the conflicting sides in any significant dispute, and provide both sides with roughly equal attention' (Boykoff and Boykoff 2007, p. 1193, citing Entman 1989, p. 30). Indeed, from professional articles or news stories, we would like to see the information from the different sides in order to make up our own minds.

In the case of scientific topics, such as climate change, it is much more complicated: 'simply to balance sides gives readers little guidance about the scientific significance of different views' (Nelkin 1995, p. 88). Should the problem be presented equally from the point of view of the scientists confirming the anthropogenic character of climate change as well as from the point of view of so-called climate sceptics who either reject the whole idea of climate change or the human impact on it? As Carvalho (2007, p. 223) points out, 'media depiction of the issues often suggest that the scientific community is divided in the middle'. This problem partly concerns the understanding of what 'balance' means. Is it according equal weight to two unequal sides, or is it according *proportionate* weight to each side? For instance, Anderegg et al. (2010, p. 1) state that '97–98 percent of the climate researchers most actively publishing in the field support tenets of anthropogenic climate change', so if journalists quote an equal number of scientists from both sides of the debate and prescribe equal weight to their arguments, does it represent balanced reporting of the problem? On the contrary, this journalistic quest for 'objectivity' leads to a 'distortion of the news' (Antilla 2005, p. 339) and consequently to public belief in the greater divide between scientists (Freudenburg and Muselli 2010).

The damaging character of 'balanced' media coverage of climate change might lead to climate change scepticism amongst the public (Liu et al. 2008) and more importantly might negatively affect national (or even international) policy on climate change by giving policy makers a justification not to act. For example, as Fletcher (2009) states, the controversy created around climate change diverted attention from the Bush administration's climate change policy. Antilla (2005) supports the argument established by Zehr and Gelbspan, where they explain how journalists' professional standards, such as objectivity, make them present the opinions of 'industry-supported science'[6] to balance their coverage (Durfee and Corbett 2005, p. 88).

Recently, in the United States and Europe the situation has been changing, and even Boykoff (2007, p. 470) started to question whether the norm of balance in climate change media coverage still creates problems or is a 'dead norm'. His later comparative research of US and UK newspapers has demonstrated 'a dramatic increase in the quantity of newspaper coverage of anthropogenic climate change' (ibid, p. 475). Ward (2008) confirms that since 2006 in the USA more and more coverage is devoted to the scientific consensus on climate change, rather than its sceptics. Doulton and Brown (2007) have also challenged Boykoff and Boykoff's (2004) original research by stating that their media analysis of UK newspapers did not demonstrate a 'balanced' approach to the coverage, but rather the predominant majority of articles were quite alarming and urged an active climate change policy. Media analysis of newspapers from New Zealand and Australia has also demonstrated that media discourse has moved from questioning the science to finding a solution (Howard-Williams 2009). The same conclusions were made by Grundmann and Scott (2014) in their comparative study among the UK, the USA, Germany and France. This change towards 'unbalanced' reporting is a positive move in the media coverage of climate change, as Ereaut and Segnit (2006, p. 25) state that 'treating climate change as beyond argument' is one of the greatest steps on the way to popularising the climate change topic and approaching proactive mitigation policy.

In summary, one of the most popular approaches to the studies on media coverage of climate change introduced by Boykoff and Boykoff argues that the media failure to accurately popularise scientific findings in the field is not accidental but rather is a result of systematic 'micro-processes' (Boykoff and Boykoff 2004, p. 134) stimulated by journalists' professional norms. Though later researchers have shown that the journalistic norm of balance is losing its relevance, overall the influence of journalism's nature cannot be underestimated. For the purpose of this book, 'micro-processes' or 'micro-factors' are understood as the variables which are specific to journalists' reporting of climate change, such as journalists' norms or the scientific specifications of the topic, whilst 'macro-processes' or 'macro-factors' are seen as the politico-economic and social context within which journalists have to operate, such as the influence of capitalist ideology or state policy on climate change (as discussed below).

Does politics matter in covering climate change?

Throughout the last two decades, we have witnessed how the science of climate change has been replaced by the politics of climate change. The shift from scientific sources of information to political ones started in the late 1980s and continues today (Bell 1994, Wilkins 1993). And especially after a scientific consensus (or close to it) was reached (mostly through the IPCC reports), public eyes turned to the politicians as they tried to negotiate on greenhouse gas reduction commitments. The goal of trying to keep the temperature rise below 2°C is considered to have a political rather than a scientific base. Various studies have demonstrated the nuances and complexity of the international negotiations around climate change which detract attention from the scientific research on the subject matter (Doulton and Brown 2007, Schafer et al. 2014) and that the influence of politics 'complicates efforts to move ahead with any kind of consensus or compromise on climate change despite the urgency of the issue' (Kim 2011, p. 691). Even scientific uncertainty, as discussed above, becomes 'a powerful political tool' (Boykoff and Boykoff 2007, p. 1193), where any hesitation in scientific agreements is used in order to postpone or twist the political decisions. At the same time, the popularity of climate change as a news item has grown proportionately to the influence of politics. It has been demonstrated that media coverage is particularly amplified during major international events (Gavin and Marshal 2011, Mercado 2012), which have 'significant attention-grabbing power' (Liu et al. 2011, p. 415). In media coverage, the voice of political actors is more prominent than those of any other actors, including scientists, businesses or environmental activists.

Carvalho and Burgess (2005, p. 1457), in their analysis of UK broadsheet newspapers' coverage of climate change, argue that the discourse created by the media around this topic is influenced by 'the agency of top political figures and the dominant ideological standpoints in different newspapers.' The media's stance on the certainty of climate change or the role of the state in the climate change mitigation process depends on the newspapers' political affiliation. For example, when the first IPCC report was announced in 1990, the conservative paper *The Times* published a series of articles seeking to discredit the scientific discoveries on climate change risks, whilst the more liberal *Guardian* accentuated the 'danger' of climate change and used it as an opportunity to criticise the government and its official position. Grundmann and Scott (2014, p. 233) also argue the important connection between media attention to the climate change problem and the stance of the 'political elites and their agreement or disagreement'.

What is even more interesting, as Carvalho (2005, p. 21) points out in the case of the UK quality press, is that despite the difference between the newspapers' political orientation and its impact on climate change coverage, most of the time they all 'remain within the broad ideological parameters of free-market capitalism and neo-liberalism, avoiding a sustained critique of the possibility of constant economic growth and increasing consumption, and of the profound international injustices associated with the greenhouse effect'. Even though in some instances economic instruments and technological solutions are actively popularised by the media, as Davidsen and Graham (2014, p. 164) have observed in the case of Canadian media, this 'may divert potential debate from problems in the larger policy framework that the Canadian energy economy is embedded in'. Hence, the media discourse around climate change is not only influenced by the political affiliations of a single studied newspaper, but in general it is to some degree unified by the existing overarching state ideology. This observation is extremely important for the study of media coverage of climate change and for the fate of climate change mitigation policy in general, since arguably climate change is a product of 'a consumer society [. . .] and a "buy now, think later logic"' (Boyce and Lewis 2009, p. 5; see also Trumbo and Shanahan 2000, Wilkins 1993). Thus in order to be able to stop or slow down the tragic consequences of climate change, the whole concept of economic growth and consumerism has to be modified.

Another approach to the study of the interrelations between media coverage of climate change and state politics has been suggested by Neil Gavin (2009), who draws our attention to the argument that as much as politics shapes the media, the media can motivate politicians on certain behaviour as well. In other words, we can witness the process of the 'mediatisation' of politics (Gavin 2009 citing Meyer 2002). The idea of 'mediatisation of politics' refers to the media's power over politicians' behaviour through the 'third person effect', which states that even if a person was not persuaded by the news, he/she might think that others will be. For instance, even if politicians do not have evidence of media influence over public opinion on the climate change topic, they might think that this influence exists, which in turn makes them attentive to how mass media present such problems as climate change. Climate change mitigation policy assumes 'unpopular political effects in many countries, with the need to reduce power consumption, vehicle use and other "everyday luxuries" which industrialised societies take for granted' (Bell 1994, p. 59). Therefore 'political self-preservation' (Gavin 2009, p. 168) makes politicians very cautious if 'intervention [into climate change policy. . .] involves direct or financially burdensome initiatives, awkward and intrusive regulatory policies, or higher taxation' (ibid, p. 768). Eventually, one of the conclusions Gavin (2009, p. 771) reached in his research was that in the UK context, even if they wanted to, politicians would not be able 'to push climate change further up the media agenda', but rather they have to think about themselves and how to react to certain media messages. For instance, during the UK parliamentary election campaign of 2010, all major parties were consistently referring to environmental issues (including climate change) – however, closer to the day of elections, priority was still given to the more publicly discussed problems of economic crisis and immigration (Rootes and Carter 2010). Doulton and Brown (2007) also argue that if government (in the UK) wants to take leadership in the international action on climate change, it needs to be sure that its actions will be backed by public support, and therefore it is important to understand what information on climate change the public receives.

In conclusion, politicisation of the problem and bringing the national agenda into media coverage of climate change might be considered a powerful tool in attracting attention to climate change. In the case of Russia, as will be seen throughout this study, we face a different set of problems, even though the politicisation of climate change reporting is relevant to it. If we are talking about the influence of the state on the media system on climate change discourse, we should take into consideration that there is a different type of regime, a different understanding of climate change problems and consequently a different framing of the problem. The concluding part of this chapter will discuss the existing studies on media coverage of climate change in Russia and the theoretical framework used in this study.

Russian media, climate change and the Propaganda Model

Studies of the media and climate change in Russia were virtually non-existent until a few years ago, and still their number remains extremely limited. To date, only a few authors working on this topic can be identified. One of them, Dmitry Yagodin (2010), in his brief analysis of the media coverage of the Copenhagen Conference in 2009 in two newspapers – the tabloid *Moskovsky Komsomolets* and the quality paper *Kommersant*, highlights how newspapers mostly follow state official policies on climate change, and the state in turn backs up the interests of large industries. Nina Tynkkynen (2010, p. 182) draws a connection between the Russian print media's framing of climate change and the concept of Russia as a 'Great Ecological Power'. She relates this idea to the historical concept of Russia considering itself a 'Great Power', which she traces back to the sixteenth century and connects it with the modern political situation, such as a restoration of Russia's 'greatness' under the presidency of Putin.

Elana Wilson Rowe (2009) studies the coverage of climate change in the state-owned newspaper *Rossiyskaya gazeta* in order to analyse the role of experts in the framing of this issue. The author acknowledges that it is quite difficult to assess how much the expert community influences Russia's policy on climate change; however, Wilson Rowe (2009, p. 607) concludes that 'their [experts'] presence appears to be deemed necessary and appropriate' and scientists see themselves 'as having policy-related roles' (ibid, p. 608). As probably could be expected, the role of scientists becomes more prominent when their position coincides with that of the state. Wilson Rowe (2009) concludes that Russia will stay engaged with the international negotiations on climate change, as it does not really contradict its political and economic agendas and, as Wilson Rowe argues, Russia's approach to climate change gets closer to the European one.

Even this limited amount of information on the media coverage of climate change in Russia demonstrates some degree of state influence and overall consent within media coverage. Acknowledging a vast range of potential theoretical frameworks, this book adopts Herman and Chomsky's Propaganda Model (PrM), which, as I argue, allows us to understand these complex connections between media and the state in Russia and perhaps illuminate other actors involved in the process of climate change coverage.

Edward Herman and Noam Chomsky (1994 [1988]) caused significant debate within the academic community when they claimed that American mass media do not act as the 'watchdog' for the liberal ideas which the United States is supposed to be based on. Instead they claimed that mass media was just a 'tool' in the hands of economic and political elites. Whilst in its communist adversaries (such as the USSR and the People's Republic of China), mass media were straightforwardly governed by the leading party through officially accepted censorship and institutionalised control, in the liberal democracy of the United States, as Chomsky (1989b, p. 19) said, 'more subtle' means were required to control and influence mass media. It is not very easy to find a Western journalist who would admit that he/she has experienced open censorship by the ruling elites or that words are put into his/her mouth. Indeed this is not what Herman and Chomsky stated in their work. On the contrary, they argued that the political economy of mass media is such that it becomes quite natural for the reporters to cover a news story in one way or another.

Herman and Chomsky explained their vision of how mass media operates by introducing the PrM, which suggests that before information reaches the pages of newspapers or is broadcast on television, it goes through five filters, or barriers: the media ownership structure (usually, mass media sources belong to a large corporation and have to support its interests), advertising (to cover the costs of production and make more profit, the media try to find a way to match their content to advertisers' requirements), information sources (journalists' dependence on newsmakers), flak (negative reaction to media activity; it appears after media publish or broadcast information which is not in favour of some individual or political or business groups) and anti-communist ideology (after the Soviet Union fell, it was replaced by other anti-ideologies, such as anti-terrorism).

The foundation of the PrM rests on the idea of media compliance with governmental and/or economic elites and downplays the liberal-pluralist view of the function of the media, which claims that 'the media serve as a guardian of the public interest and as a watchdog on the exercise of power' (Mullen 2010b, p. 674). The PrM's arguments on questioning the role of the media in the society are arguably close to the ones proposed by the followers of the Marxist-radical tradition (Mullen and Klaehn 2010), who see the media as a tool in the hands of the dominating classes, used in order to achieve certain political goals. In this case the goal is the 'continuation of the capitalist class system' through news content (Pedro 2011a, p. 1866). From another point of view, Herman and Chomsky's approach to understanding the media production process belongs to the tradition of the political economy of communication studies (Mullen and Klaehn 2010), which looks at the media in the wider politico-societal context (see also McChesney 1998).

To some extent, the PrM resembles the Marxist vision of the media; Chomsky (1989a) himself stresses that 'the similarity between this [the ideas expressed by Lasswell and Lippmann] and Leninist ideology is very striking. According to Leninist ideology, the cool observers, the radical intelligentsia, will be the vanguard who will lead the stupid and ignorant masses on to communist utopias, because they are too stupid to work it out by themselves' (ibid, p. 6). Even though Herman and Chomsky do not directly situate their work within the Marxist or

Leninist traditions, neither do they use any Marxist terminology; but considering the influence and references (even in the title of their book) to the work of Lasswell and Lippmann, Herman and Chomsky indirectly relate and compare their approach to the one developed earlier by Marxists. It can be argued that the PrM does indeed resemble Leninism, but not with its paternalism, rather with Lenin's scepticism about the bourgeois press. Such as Lenin saw the bourgeois press as a source of mass deception which through its lies tried to slander the Bolsheviks. Furthermore, he argued that despite the possibility of the formal absence of censorship, bourgeois freedom of the press is a 'freedom of the rich, the bourgeoisie, to deceive the oppressed and exploited masses' (Resis 1977, p. 282, citing Lenin 1917, p. 209), whilst genuine freedom of the press suggests that the 'press must be liberated from the power of money as well as from the power of the censor' (Resis 1977, p. 282).

The PrM suggests 'an institutional critique of mass media' (Klaehn 2002, p. 170). It demonstrates the media's dependence on sources of funds and power and argues that everyday media practices are bounded by propaganda and that media are forced to fit around the interests of the elite. It gives a broad framework of analysis for a very complicated system of social events, so it needs to be modified from case to case. When adapting the PrM to the Russian case of media coverage of climate change, it is important to take into consideration the country's political, economic and social characteristics and be prepared to adjust the PrM filters where necessary. Overall it is hoped that the application of the PrM to this case study will not only theoretically benefit the body of media studies literature, but also (as Herman and Chomsky idealistically suggested) provide an opportunity for environmentalists and grassroots movements to see their way around the system and perhaps to some degree change it.

It should be noted that Herman and Chomsky's model was heavily criticised in academic literature for its determinism and simplicity, often dismissed on the grounds of its conspiratorial nature and ignored in mainstream media studies in the United States (Jensen 2010, Mullen 2010a, Robertson 2010). However, the authors kept arguing for the adequacy and high value of their approach to understanding the media system in general and of the PrM in particular, and continued to develop their ideas in a number of other studies (for example, see Chomsky 1989a and 1989b; Herman 1999 and 2000; Herman and Chomsky 2002; Herman and McChesney 1997). The PrM was also discussed by numerous dedicated researchers (Jensen 2010, Klaehn 2003, 2005, 2009a and 2009b; McChesney 1998, Mullen 2008, 2010a and 2010b; Mullen and Klaehn 2010, Pedro 2011a and 2011b).

The ideas populated by Herman and Chomsky can also be found in a number of academic articles dedicated to media coverage of climate change. For example, Carvalho's (2005) research has demonstrated how the UK broadsheet newspapers differ in their ideological orientation, which made the author state that 'factors like ownership and the wider political economy of the media can provide significant contribution to understanding these differences, as well as the press's relations with established interests and the social distribution of power' (ibid, p. 21). As an example, Carvalho referred to Herman and Chomsky's PrM, which relies on the idea that media coverage is influenced by the political and economic context. Olausson (2009) also refers to the studies (including Herman and Chomsky's) on the media's conformist position to official policy and confirms the relevance of the argument with regard to the media coverage of climate change, on not only the national but also the international scale of Europe. Eventually she concludes that 'the tight relationship between the political elite and the media implies that the media do not offer any alternative frames, in relation to those established in policy discourse, for understanding global climate change' (ibid, p. 433).

Holmes (2009, p. 99) does not directly refer to the PrM; however, he points out 'the surrounding framework of institutional pressures from owners, managers, and major sources of revenue; the capacity of various sources to mobilise "flak"; and propaganda campaign, PR, and information management' (ibid), which make a substantial contribution to the shaping of the journalistic norms. Wu (2009, p. 165) also alludes indirectly to some postulates of the PrM by stressing possible dependency between the media's economic and political stance and their approach to climate change coverage.

Good (2008) and Babe (2005) specifically apply the PrM to the analysis of the coverage of climate change. Babe (2005), through an analysis of the media coverage of global warming and the Kyoto Protocol in Canadian newspapers, concluded that the coverage was 'consistent with the PrM[. . . .] [Environmental] issues were never addressed in their full range and seriousness; lip-service, we might speculate, served to divert attention from the overall thrust of the reporting, which was one-sided and hardly environmental' (ibid, p. 219). Good (2008, p. 234) justifies her choice of research topic and the theoretical approach by saying of climate change that 'there is arguably no other issue that is on the one hand so fundamentally challenging to the interests of the global elite neo-liberal order, and yet has consequences that are so easily framed, or ignored as something else'. Consequently, the author argued that the American media frame climate change issues in a way that corresponds to the elite's interests – for instance, 'to avoid critique of the world's largest, and most profitable, industry: oil' (ibid, p. 235). Such a specific conclusion derives from the analysis of the US role in climate change policy and where its position lies.

Concluding remarks

As the foregoing analysis has demonstrated, media reporting on climate change is a complex process which involves consideration of macro-factors (such as politics and economy) and micro-processes (for example, the widely discussed journalistic norms and practices). It seems that the latter arguments have been explored much more in the bulk of the literature, while few have investigated whether these micro-processes could not be enough for understanding why climate change in Russia is covered or not covered in a certain way. As Carvalho (2007, p. 225) states, 'the role of ideology in media representation of science is still blatantly under-researched'; in the Russian case, the role of ideology and the state and the way it interacts with other actors influencing media reporting of climate change is not researched at all. This book will allow us to understand not only the way the media and the state coordinate with each other in Russia, but also the political discourse that has evolved around climate change issues and the role of other actors involved in this process. This goal will be achieved by deconstructing the Russian media system into the elements of the PrM and analysing how these macro-factors shape media coverage of climate change in the country.

Notes

1 By 'risks' Beck (1994, p. 22) understands not only 'all radioactivity, which completely evades human perceptive abilities, but also toxins and pollutants in the air, the water and foodstuffs, together with the accompanying short- and long-term effects on plants, animals and people.'
2 Shanahan (2009) states that in non-industrialised countries the problem of the lack of special training on how to write about climate change is also worsened by the journalists' inability to find local experts on the topic who will communicate their findings, to persuade editors about the importance of climate change and to have resources to travel to collect information or attend conferences.
3 These examples are taken from the Russian media.
4 As an example, Boykoff (2011) points to the situation after the 'Climategate' scandal, when scientists did their best to avoid any form of communication with journalists. For more on issues of mistrust between journalists and scientists, see Nelkin (1995).
5 Nissani (1999) says that what readers get is an enormous amount of politicians' or businessmen's presentations, reports from international and national meetings and negotiations and scientists' opinions. The information coming from these sources presupposes that people already have knowledge about the topic and can put it into context without external help, but according to Nissani, most people do not have the necessary background, and information in newspapers is useful only to environmental specialists.
6 One of the examples: 'a coalition headed by the American Petroleum Institute invested $600,000 in 1998 in a campaign aiming to increase the volume of US news coverage questioning the prevailing climate science' (Doulton and Brown 2007, p. 2, citing Cushman 1998).

Bibliography

Akpan, C., Anorue, L. and Ukonu, M. (2012) 'An Analysis of the Influence of the Nigerian Mass Media on Public Understanding of Climate Change', *Journal of Alternative Perspectives in the Social Sciences*, 4/4: 688–710.

Anderegg, W. Prall, J., Harold, J. and Schneidera, S. (2010) 'Expert Credibility in Climate Change', *PNAS*, 107/27: 12107–12109.

Antilla, L. (2005) 'Climate of Scepticism: US Newspaper Coverage of the Science of Climate Change', *Global Environmental Change*, 15: 338–352.

Babe, R. (2005) 'Newspaper Discourses on Environment'. In: J. Klaehn (ed) *Filtering the News. Essays on Herman and Chomsky's Propaganda Model*, London: Black Rose Books: 187–222.

Beck, U. (1994) *Risk Society. Towards New Modernity*, London: Sage.

Bell, A. (1994) 'Climates of Opinion: Public and Media Discourse on the Global Environment', *Discourse and Society*, 5/1: 33–64.

Bersenev, E. (2009) 'Zimniy Privet ot Global'nogo Potepleniya', *Sibnet.ru* (18 February), http://info.sibnet.ru/?id=265980, date accessed 29/08/2014.

Bostrom, A. and Lashof, D. (2007) 'Weather or Climate Change?'. In: S.C. Moser and L. Dilling (eds.) *Creating a Climate for Change: Communication Climate Change and Facilitating Social Change*, Cambridge: Cambridge University Press.

Boyce, T. and Lewis, J. (eds.) (2009) *Climate Change and the Media*, New York: Peter Lang Publishing.

Boykoff, J. (2012) 'US Media Coverage of the Cancun Climate Change Conference', *PS: Political Science & Politics*, 45: 251–258.

Boykoff, M. (2007) 'Flogging a Dead Norm: Newspaper Coverage of Anthropogenic Climate Change in the United States and United Kingdom from 2003 to 2006', *Area*, 39/4: 470–481.

Boykoff, M. (2008a) 'The Cultural Politics of Climate Change Discourse in UK Tabloids', *Political Geography*, 27: 549–569.

Boykoff, M. (2008b) 'Lost in Translation? United States Television News Coverage of Anthropogenic Climate Change, 1995–2004', *Climatic Change*, 86: 1–11.

Boykoff, M. (2011) *Who Speaks for the Climate? Making Sense of Media Reporting on Climate Change*, Cambridge: Cambridge University Press.

Boykoff, M. and Boykoff, J. (2004) 'Balance as Bias: Global Warming and the US Prestige Press', *Global Environmental Change*, 14: 125–136.

Boykoff, M. and Boykoff, J. (2007) 'Climate Change and Journalistic Norms: A Case-study of US Mass Media Coverage', *Geoforum*, 38/6: 1190–1204.

Busygin, A. (2005) 'Global'noe Poteplenie Neset v Evropu Chumnuyu Zarazu', *Utro.ru*, www.utro.ru/articles/2005/11/14/495017.shtml, date accessed 12/12/2010.

Butler, C. and Pigeon, N. (2009) 'Media Communication and Public Understanding of Climate Change: Reporting Scientific Consensus on Anthropogenic Warming'. In: T. Boyce and J. Lewis (eds.) *Climate Change and the Media*, New York: Peter Lang Publishing: 43–58.

Carvalho, A. (2005) 'Representing the Politics of the Greenhouse Effect: Discursive Strategies in the British Media', *Critical Discourse Studies*, 2/1: 1–29.

Carvalho, A. (2007) 'Ideological Cultures and Media Discourses on Scientific Knowledge: Re-reading News on Climate Change', *Public Understanding of Science*, 16: 223–243.

Carvalho, A. and Burgess, J. (2005) 'Cultural Circuits of Climate Change in UK Broadsheet Newspapers, 1985–2003', *Risk Analysis*, 25/6: 1457–1469.

Chomsky, N. (1989a) 'Manufacturing Consent: The Political Economy of the Mass Media'; talk delivered at the University of Wisconsin, Madison, 15 March. www.chomsky.info/talks/19890315.htm, date accessed 9/3/2015.

Chomsky, N. (1989b) *Necessary Illusions. Thought Control in Democratic Societies*, London: Pluto Press.

Davidsen, C. and Graham, D. (2014) 'Newspaper Reporting on Climate Change, Green Energy and Carbon Reduction Strategies across Canada 1999–2009', *American Review of Canadian Studies*, 44/2: 151–168.

Devine, J. (2012) 'Is There a Strong Link between Extreme Weather Events and Climate Change?' *Scientific American* (30 September), www.scientificamerican.com/article.cfm?id=climate-change-and-extreme-weather, date accessed 06/11/2012.

Dirikx, A. and Gelders, D. (2009) 'Global Warming through the Same Lens: An Explorative Framing Study in Dutch and French Newspapers'. In: T. Boyce and J. Lewis (eds.) *Climate Change and the Media*, New York: Peter Lang Publishing: 200–210.

Doulton, H. and Brown, K. (2007) 'Ten Years to Prevent Catastrophe'? Discourse of Climate Change and International Development in the UK Press', *Tyndall Centre Working Paper* 111: 1–58.

Downs, A. (1972) 'Up and Down with Ecology – the "Issue-Attention Cycle"'. *The Public Interest*, 28: 38–50.

Dunwoody, S. and Griffin, R. (2002) 'Judgmental Heuristics and News Reporting'. In: R. Gowda and J. Fox (eds.) *Judgements, Decisions, and Public Policy*, Cambridge: Cambridge University Press: 177–198.

Durfee, J. and Corbett, J. (2005) 'Context and Controversy: Global Warming Coverage', *Nieman Reports*, 59/4: 88–89.

Ereaut, G. and Segnit, N. (2006) 'Warm Words: How Are We Telling the Climate Story and Can We Tell It Better? Institute for Public Research', www.ippr.org, date accessed 11/11/2009.

Fletcher, A. (2009) 'Clearing the Air: the Contribution of Frame Analysis to Understanding Climate Policy in the United States', *Environmental Politics*, 18/5: 800–816.

Freudenburg, W. and Muselli, V. (2010) 'Global Warming Estimates, Media Expectations, and the Asymmetry of Scientific Challenge', *Global Environmental Change*, 20: 483–491.

Gavin, N. (2009) 'Addressing Climate Change: A Media Perspective', *Environmental Politics*, 18/5: 765–780.

Gavin, N. and Marshall, T. (2011) 'Mediated Climate Change in Britain: Scepticism on the Web and on Television around Copenhagen', *Global Environmental Change*, 21/1: 1035–1044.

Gavin, N., Leonard-Milsom, L. and Montgomery, J. (2011) 'Climate Change, Flooding and the Media in Britain', *Public Understanding of Science*, 20/3: 422–438.

Good, J. (2008) 'The Framing of Climate Change in Canadian, American, and International Newspapers: A Media Propaganda Model Analysis', *Canadian Journal of Communication*, 33: 233–255.

Grundmann, R. and Scott, M. (2014) 'Disputed Climate Science in the Media: Do Countries Matter?', *Public Understanding of Science*, 23/2: 220–235.

Hansson, S. O. (2004) 'Fallacies of Risk', *Journal of Risk Research*, 7/3: 353–360.

Herman, E. (1999) *The Myth of the Liberal Media*, New York: Peter Lang.

Herman, E. (2000) 'The Propaganda Model: A Retrospective', *Journalism Studies*, 1/1: 101–112.

Herman, E. and Chomsky, N. (1994 [1988]) *Manufacturing Consent. The Political Economy of the Mass Media*, London: Vintage.

Herman, E. and Chomsky, N. (2002) *Manufacturing Consent. The Political Economy of the Mass Media. With a New Introduction by the Authors*, New York: Pantheon Books.

Herman, E. and McChesney, R. (1997) *The Global Media. The New Missionaries of Global Capitalism*, London: Cassel.

Hoijer, B. (2010) 'Emotional Anchoring and Objectification in Media Reporting on Climate Change', *Public Understanding of Science*, 19/6: 717–731.

Holmes, T. (2009) 'Balancing Acts: PR, "Impartiality" and Power in Mass Media Coverage of Climate Change'. In: T. Boyce and J. Lewis (eds.) *Climate Change and the Media*, New York: Peter Lang Publishing: 92–100.

Howard-Williams, R. (2009) 'Ideological Construction of Climate Change in Australian and New Zealand Newspapers'. In: T. Boyce and J. Lewis (eds.) *Climate Change and the Media*, New York: Peter Lang Publishing: 28–40.

Hulme, M. (2009) 'Mediated Messages about Climate Change: Reporting the IPCC Fourth Assessment in the UK Print Media'. In: T. Boyce and J. Lewis (eds.) *Climate Change and the Media*, New York: Peter Lang Publishing: 117–128.

Jensen, R. (2010) 'The Faculty Filter: Why the Propaganda Model Is Marginalized in US Journalism Schools'. In: J. Klaehn (ed.) *The Political Economy of Media and Power*, New York: Peter Lang Publishing: 235–242.

Karavaev, N. (2005) 'Chelovechestvo ne Perezhivet XXI Vek', *Utro.ru*, www.utro.ru/articles/2005/09/30/481882.shtml, date accessed 12/12/2010.

Keeling, S. (2009) 'What If? Media, Celebrity and Climate Change', *Weather*, 64/2: 49–50.

Kim, K.S. (2011) 'Public Understanding of the Politics of Global Warming in the News Media: The Hostile Media Approach', *Public Understanding of Science*, 20/5: 690–705.

Klaehn, J. (2002) 'A Critical Review and Assessment of Herman and Chomsky's "Propaganda Model"', *European Journal of Communication*, 17/2: 147–182.

Klaehn, J. (2003) 'Behind the Invisible Curtain of Scholarly Criticism: Revisiting the Propaganda Model', *Journalism Studies*, 4/3: 359–369.

Klaehn, J. (2005) 'A Critical Review and Assessment of Herman and Chomsky's "Propaganda Model"'. In: J. Klaehn (ed.) *Filtering the News. Essays on Herman and Chomsky's Propaganda Model*, London: Black Rose Books: 1–20.

Klaehn, J. (2009a) 'Comment: Media, Power and the Origins of the Propaganda Model: An Interview with Edward S. Herman', *Fifth-Estate-Online – International Journal of Radical Mass Media Criticism*, www.fifth-estate-online.co.uk/comment/mediapower.html, date accessed 02/04/2010.

Klaehn, J. (2009b) 'The Propaganda Model: Theoretical and Methodological Considerations', *Westminster Papers in Communication and Culture*, 6/2: 43–58.

Latynina, U. (2010) 'Net Global'nogo Potepleniya – Est' Global'naya Byurokratiya', *Komsomol'skaya pravda* (18 February), www.kp.ru/daily/24444.3/608895/, date accessed 29/08/2014.

Liu, X., Vedlitz, A. and Alston, L. (2008) 'Regional News Portrayals of Global Warming and Climate Change', *Environmental Science and Policy*, 11: 379–393.

Liu, X., Lindquist, E. and Vedlitz, A. (2011) 'Explaining Media and Congressional Attention to Global Climate Change, 1969–2005: An Empirical Test of Agenda-Setting Theory', *Political Research Quarterly*, 64/2: 405–419.

Lockwood, A. (2009) 'Preparations for a Post-Kyoto Media Coverage of UK Climate Policy'. In: T. Boyce and J. Lewis (eds.) *Climate Change and the Media*, New York: Peter Lang Publishing: 186–199.

Lyytimäki, J. (2011). 'Mainstreaming Climate Policy: the Role of Media Coverage in Finland', *Mitigation and Adaption Strategies for Global Change*, 16: 649–661.

McChesney, R. (1998) 'The Political Economy of Global Communication'. In: R. McChesney, E.M. Wood and J.B. Foster (eds.) *The Political Economy of the Global Communication Revolution*, New York: Monthly Review Press: 1–26.

McComas, K. and Shanahan, J. (1999) 'Telling Stories about Global Climate Change: Measuring the Impact of Narratives on Issue Cycles', *Communication Research* 26/1: 30–57.

McManus, P. (2000) 'Beyond Kyoto? Media Representation of an Environmental Issue', *Australian Geographical Studies*, 38/3: 306–319.

Mercado, M.T. (2012). 'Media Representations of Climate Change in the Argentinean Press', *Journalism Studies*, 13/2: 193–209.

Mullen, A. (2008) 'Comment: Twenty Years at the Margins: The Herman-Chomsky Propaganda Model, 1988–2008', *Fifth-Estate-Online – International Journal of Radical*

Mass Media Criticism, www.fifth-estate-online.co.uk/comment/twentyyears.html, date accessed 24/03/2010.

Mullen, A. (2010a) 'Bringing Power Back In: The Herman-Chomsky Propaganda Model, 1988–2008'. In: J. Klaehn (ed.) *The Political Economy of Media and Power*, New York: Peter Lang Publishing: 207–234.

Mullen, A. (2010b) 'Twenty Years On: The Second-order Prediction of the Herman-Chomsky Propaganda Model', *Media, Culture and Society*, 32/4: 673–690.

Mullen, A. and Klaehn, J. (2010) 'The Herman-Chomsky Propaganda Model: A Critical Approach to Analysing Mass Media Behaviour', *Sociology Compass*, 4/4: 215–229.

Nelkin, D. (1995) *Selling Science. How the Press Covers Science and Technology*, revised ed., New York: W. H. Freeman and Company.

Nielsen, K. H. and Kjaergaard, R. S. (2011) 'News Coverage of Climate Change in Nature News and Science Now during 2007', *Environmental Communication*, 5/1: 25–44.

Nissani, M. (1999) 'Media Coverage of the Greenhouse Effect', *Population and Environment: A Journal of Interdisciplinary Studies*, 21/1: 27–43.

Olausson, U. (2009) 'Global Warming – Global Responsibility? Media Frames of Collective Action and Scientific Certainty', *Public Understanding of Science*, 18/4: 421–436.

Painter, J. (2013) *Climate Change in the Media: Reporting Risk and Uncertainty*, London: I. B. Tauris.

Pasquare, F. and Oppizzi, P. (2012) 'How Do the Media Affect Public Perception of Climate Change and Geohazards? An Italian Case Study', *Global and Planetary Change*, 90–91: 152–157.

Pedro, J. (2011a) 'The Propaganda Model in the Early 21st Century. Part 1', *International Journal of Communication*, 5: 1865–1905.

Pedro, J. (2011b) 'The Propaganda Model in the Early 21st Century. Part 2', *International Journal of Communication*, 5: 1906–1926.

Pralle, S. (2009) 'Agenda-Setting and Climate Change', *Environmental Politics*, 18/5: 781–799.

Resis, A. (1977) 'Lenin on Freedom of the Press', *Russian Review*, 36/3: 274–296.

Rhaman, M. (2010) 'Bangladesh: A Metaphor for the World'. In: E. Eide, R. Kunelius and V. Kumpu (eds.) *Global Climate – Local Journalisms. A Transnational Study of How Media Make Sense of Climate Summits*, Bochum: Projektverlag: 67–81.

Robertson, J. (2010) 'The Propaganda Model in 2011: Stronger Yet Still Neglected in UK Higher Education?' *Synæsthesia*, 1/1: 24–33.

Rootes, C. and Carter, N. (2010) 'Take Blue, Add Yellow, Get Green? The Environment in the UK General Election of 6 May 2010', *Environmental Politics*, 19/6: 992–999.

Sampei, Y. and Aoyagi-Usui, M. (2009) 'Mass-Media Coverage, Its Influence on Public Awareness of Climate Change Issues, and Implications for Japan's National Campaign to Reduce Greenhouse Gas Emissions', *Global Environmental Change*, 19: 203–212.

Schäfer, M., Ivanova, A. and Schmidt, A. (2014) 'What Drives Media Attention for Climate Change? Explaining Issue Attention in Australian, German and Indian Print Media from 1996 to 2010', *International Communication Gazette*, 76/2: 152–176.

Schmidt, A., Ivanova, A. and Schafer, M. (2013) 'Media Attention for Climate Change around the World: A Comparative Analysis of Newspaper Coverage in 27 Countries', *Global Environmental Change*, 23: 1233–1248.

Schumpeter, J. (1943) *Capitalism, Socialism and Democracy*, London: Allen & Unwin.

Shanahan, J. and Good, J. (2000) 'Heat and Hot Air: Influence of Local Temperature on Journalists' Coverage of Global Warming', *Public Understanding of Science*, 9/3: 285–295.

Shanahan, M. (2007) 'Talking about a Revolution: Climate Change and the Media', *IIED Briefing*, www.iied.org, date accessed 15/12/2009.

Shanahan, M. (2009) 'Time to Adapt? Media Coverage of Climate Change in Non-industrialized Countries'. In: T. Boyce and J. Lewis (eds.) *Climate Change and the Media*, New York: Peter Lang Publishing: 145–157.

Smith, J. (2005) 'Dangerous News: Media Decision Making about Climate Change Risk', *Risk Analysis*, 25/6: 1471–1482.

Spence, A., Poortinga, W., Butler, C. and Pidgeon, N. F. (2011) 'Perceptions of Climate Change and Willingness to Save Energy Related to Flood Experience', *Nature Climate Change*, 1: 46–49.

Stocking, H. and Holstein, L. (2009) 'Manufacturing Doubt: Journalists' Roles and the Construction of Ignorance in a Scientific Controversy', *Public Understanding of Science*, 18/1: 23–42.

Sunstein, C. (2006) 'The Availability Heuristic, Intuitive Cost-Benefit Analysis, and Climate Change', *Climate Change*, 77: 195–210.

Taddicken, M. (2013) 'Climate Change from the User's Perspective: The Impact of Mass Media and Internet Use and Individual and Moderating Variables on Knowledge and Attitudes', *Journal of Media Psychology*, 25/1: 39–52.

Taylor, N. and Nathan, S. (2002) 'How Science Contributes to Environmental Reporting in British Newspapers: A Case Study of the Reporting of Global Warming and Climate Change', *The Environmentalist*, 22: 325–331.

Trumbo, C. (1996) 'Constructing Climate Change: Claims and Frames in US News Coverage of an Environmental Issue,' *Public Understanding of Science* 5: 269–273.

Trumbo, C. and Shanahan, J. (2000). 'Social Research on Climate Change: Where We Have Been, Where We are At, and Where We Might Go', *Public Understanding of Science*, 9/3: 199–204.

Tynkkynen, N. (2010) 'A Great Ecological Power in Global Climate Policy? Framing Climate Change as a Policy Problem in Russian Public Discussion', *Environmental Politics*, 19: 179–195.

Ward, B. (2008) 'A Higher Standard than "Balance" in Journalism on Climate Change Science: An Editorial Comment', *Climatic Change*, 86: 13–17.

Wilkins, L. (1993) 'Between Facts and Values: Print Media Coverage of the Greenhouse Effect, 1987–1990', *Public Understanding of Science*, 2: 71–84.

Wilson Rowe, E. (2009) 'Who Is to Blame? Agency, Causality, Responsibility and the Role of Experts in Russian Framings of Global Climate Change', *Europe-Asia Studies*, 61/4: 593–619.

Wu, Y. (2009) 'The Good, the Bad, and the Ugly: Framing of China in News Media Coverage of Global Climate Change'. In: T. Boyce and J. Lewis (eds.) *Climate Change and the Media*, New York: Peter Lang Publishing: 158–173.

Xu, P. (2010) 'China: Emerging Player with a Historical Legacy'. In: E. Eide, R. Kunelius and V. Kumpu (eds.) *Global Climate – Local Journalisms. A Transnational Study of How Media Make Sense of Climate Summits*, Bochum: Projektverlag: 131–145.

Yagodin, D. (2010) 'Russia: Listening to the Wind – Clientelism and Climate Change'. In: E. Eide, R. Kunelius and V. Kumpu (eds.) *Global Climate – Local Journalisms. A Transnational Study of How Media Make Sense of Climate Summits*, Bochum: Projektverlag: 275–290.

Zehr, S. (2009) 'An Environmentalist/Economic Hybrid Frame in US Press Coverage of Climate Change, 2000–20008'. In: T. Boyce and J. Lewis (eds.) *Climate Change and the Media*, New York: Peter Lang Publishing: 80–91.

2 The political economy of Russian mass media

State and market

The applicability of Western models to the Russian media system has been questioned by various scholars. Sarah Oates (2007) points out that due to the interdependence of economic and political powers in Russia with regard to the media system, it becomes difficult to classify it; one may even question whether Russian media can be characterised by Western media models. In this chapter, I argue that whilst we can see some similarities between the Russian and Western media systems (especially in relation to their coverage of climate change), there are some adjustments that have to be considered. One of the suggested modifications is concerned mostly with the role of the state in the Russian media production process, which has been examined by Russian and international scholars. In the majority of the cases, the importance of the state has been confirmed. Since the time when Peter the First founded the first newspaper (*Vedomosti*), the government of the Russian Empire and then of the Soviet Union supported the popularisation of the media, but it could also be argued that Russia became an 'inventor' of 'total censorship' coming from the top[1] (Markov 2010, p. 206). For decades, Russian press freedom had been suppressed by the government to a lesser or greater degree depending on the regime or leader in power.

In the second half of the 1980s, Mikhail Gorbachev proclaimed the policy of *glasnost*,[2] and after the end of the Soviet state (and probably for the first time in history) there was hope for freedom of speech and independence of Russian media. As a result of introducing the ideology of capitalism and the free market, new actors entered the arena, and as Oates (2007, p. 1281) suggests, '[i]f the system is consumer driven, then it is much less vulnerable to manipulation, either by a powerful group of elites or by inchoate masses'. However, even in the contemporary era of free market ideology, when censorship was officially banned, the debate surrounding the degree to which Russian media can be considered free is still ongoing.

This chapter discusses the Russian media system through the prism of Herman and Chomsky's (1994) Propaganda Model filters: ownership structure, advertising, sources of information, flak and dominant ideology. Each filter will be applied to the Russian case and consideration given to how it has changed over time and what were the dominating factors which led to the current state of the Russian mass media. Based on data from a series of in-depth interviews, the analysis will be situated within the context of the discussion of climate change coverage.

Ownership structure

When talking about the barriers created by the 'ownership' filter Herman and Chomsky (1994 [1988]) warn of the monopolisation of the media market in the United States, where a few major conglomerates own and provide financial security for the major media outlets in the country and occasionally steer their information policy in the 'right' direction. The ownership structure of Russian mass media only distantly resembles the American situation – being significantly transformed over the last 25 years, it has become a hybrid where market forces do play a role but are subordinate to the state.

During the Soviet era, the CPSU was the general manager of mass media in Russia, regulating its activity through official Party legislation. The Party strictly controlled all aspects of news production processes, starting with volume, frequency, content and design and ending with the editor's relations with the audience. Media was the *propaganda tool* to achieve the aims of the Party (Strovskiy 2011). As Coyne and Leeson (2009, p. 9) state, the 'media was central to the Soviet propaganda system'. Supporting this argument, they refer to Lenin's understanding of the mass media's role in achieving the revolution and building the new societal order. Indeed, in Lenin's (1969 [1902]) famous work *Chto Delat'?* [What Is to Be Done?], he proclaimed newspapers to be a collective propagandist and collective agitator, but also a collective organiser. In his definition, Lenin compares newspapers to scaffolding, which is not part of the house, but without them you cannot build it. Voltmer (2000, p. 478) states that Lenin's vision of mass media is 'in obvious contradiction to western journalistic norms', albeit she also admits that some of his postulates are still relevant in the current situation.

Even when the time of the revolution passed, the ability of newspapers to organise people remained quite similar – 'to implement the directives and policies of the central government' (Mickiewicz 1981, p. 68). Lenin's ideas were carried on by Stalin and Khrushchev, who used/abused mass media in their own interests to impose the Soviet ideology upon the state's citizens. Throughout the Soviet era, many newspapers were unprofitable, but their financial problems were always solved by their owner – the Party. The CPSU central committee was also in charge of hiring editors and journalists, practically all of them Party workers charged with achieving the Party's goals through the means of language. Work under such conditions demanded certain behaviour from journalists. Their professional norms were restricted by discipline, acceptance of editorial decisions and fear of breaking the rules (Grabel'nikov 2001). Self-censorship is a related problem and is discussed in greater detail below.

After the fall of the USSR, the CPSU, which used to determine Soviet press policy, became part of history. Some argue that after decades of being the propaganda tool of the Soviet government, Russian mass media during *perestroika* 'turned against' its patron and played a significant role in bringing it down. Not diminishing other crucial reasons for the Soviet system's collapse, Coyne and Leeson (2009, p. 11) conclude that 'eroding economic, political, and social conditions were important factors giving media the space to create the common knowledge required to activate the tipping point necessary for this change'. The new regime and new role of media in society demanded a new type of ownership structure. To be able to speak on behalf of the whole society, critically assess the government's performance and give different perspectives on political, economic and societal events, Russian mass media ideally needed to become as independent as was possible. So, the law of mass media which was accepted in 1990 and then changed again in 1991 established the right of ownership of mass media for not only the Party, but also for non-governmental commercial organisations and even private individuals. As a result, the majority of the Party media outlets were replaced by independent press created on the basis of journalists' collectives. The vertical system of the press (from the state central newspaper *Pravda* to regional presses), which had functioned in the USSR for decades (Mickiewicz 1981, Richter 1995), was replaced by a horizontal structure which was more appropriate for the democratic principles of the new state. Unfortunately, the newspapers did not stay in the hands of journalists for long, and instead of the CPSU press, Russia got media with various ownership structures. Along with state and NGO ownership, mass media now belongs to individuals, closed joint-stock companies, open joint-stock companies, limited liability companies and so on.

The reduction of state control led to a competitive market for mass media. From that point on, media outlets have had to solve problems connected with the economic side of the media production process on their own and find ways to exist in the developing capitalist society and make a profit (Kuznetsov 2003). Simultaneously, these 'powerful actors' very quickly came to the realisation that mass media could indeed turn a profit and, perhaps more importantly, might also be a tool to realise corporate and commercial interests (Zasurskiy 2001).

The mass media laws which brought long-desired freedom also had some loopholes, which worsened the situation by not clarifying the role of the media's owners (Zasurskiy 2004), in particular how much an owner could interfere in the news production process. This ambiguity, along with other factors, led to negative consequences. The media found themselves in a situation where their owners would not openly demonstrate their influence on the news flow, pretending that the media organ was just there to inform people – up until the time when they needed to use it in their own interests (such as to conduct information wars against rivals) (Grabel'nikov 2001). Russian media once again started to lose their independence, but due to commercial reasons rather than ideological ones, this period is famous for the emergence of oligarchs into the media market (see more in Lipman and McFaul 2001). For example, at some point, Berezovsky's group owned ORT (now the First Channel), TV-6, *Kommersant*, *Nezavisimaya Gazeta*, *Novye Izvestiya*, *Ogonek* and others. Another infamous oligarch, Vladimir Gusinsky, owned NTV, NTV+, TNT, the newspaper *Segodnya*, the magazines *Itogi*, *Sem' dney*, *Karavan istoriy* and radio station *Ekho Moskvy*, whilst the newspapers *Izvestiya*, *Komsomol'skaya pravda*, *Afisha* and *Bol'shoy gorod* belonged to Vladimir Potanin. Among the obvious negative consequences of the media monopolies created by these oligarchs, Mikhail Nenashev (2010b) connects these negative events with the change in Russian mass media towards open manipulation of their audience in the interests of their owners.

Herman and Chomsky's (1994 [1988]) original vision of corporate influence over the American media industry and the monopolisation of the market can be compared to the role of oligarchs in the Russian case, who arguably represent very similar market forces. What seems to be different in the Russian situation is not so much the media's relationship to their owners, but rather the close relationship of the owners with the state. For example, today most of the above-mentioned media outlets do not exist or their ownership has changed hands. This mostly happened due to the last dramatic change in Russian media ownership structure, when Vladimir Putin first took the post of Russian president in March 2000.[3] Shortly after the start of this new era in Russian domestic politics, a series of events restricting press freedom and centralising its ownership in the hands of government started to occur (Zassoursky 2004). Oligarchs who were not in favour of the current leader eventually had to give up their media empires, which then became the property of organisations with tight state connections (governmental or private sector, or something between the two) or of some other oligarch who supported the Kremlin[4] (see more in Orttung 2006). Gusinsky's Media Holding owned the last major TV channel free of government influence, but he was forced to leave the country and his media conglomerate was sold (Pasti and Pietilainen 2008).

Currently, the number of fully independent media organs is very limited,[5] and they exist only due to their insignificance, restricted target audience or limited territorial influence – the major TV channels which cover 99 percent of Russian territory are all under government control. Ellen Mickiewicz (2008) refers to a case in 2006 when Putin, in response to the critique of media freedom in Russia, stated that according to his information the state's share in the media market was declining and the number of newspapers was growing, so it seemed impossible (to him) to control over 53,000 periodicals. Mickiewicz (2008) points out that among the impressive amount of Russian press organs mentioned by the president, there were only a very limited number which had 'a decent circulation', and the influential ones were indeed to a great extent controlled by the state or by 'clients of the government'.

Besides the vast formal state ownership of media outlets, the state also has influence through the Ministry of Culture and Mass Communication, which grants licences and publishes laws and regulations. The state also has a monopoly over transmitting equipment, such as satellites (De Smaele 2007). Coyne and Leeson (2009) argue that the negative picture of Russian media ownership structures threatens Russia's democratic development and cannot 'reinforce political and economic reforms' (ibid, p. 11). Nenashev (2010a) shares the same views on the problem by stating that the independence of the journalists' professional community is impossible due to the confrontation of the great administrative resources of the state and the financial resources of big business.

As far as the problem of the coverage of climate change goes, Putin's centralisation of power has had a long-term effect. The majority of the most important and popular media belong not only to the government and its close partners, but also to the individuals and organisations with heavy interests in the oil, gas and other industries which significantly contribute to Russia's GHG emissions. For example, Gazprom–Media Holding owns the federal TV channel NTV, the radio stations *Ekho Moskvy* and *Relax FM*, the magazines *Itogi*, *Karavan Istoriy* and *Panorama TV* and the newspaper *Tribuna* (Gazprom-Media website 2013). Alisher Usmanov, the main shareholder of Metalloinvest (one of the largest steel producers in the world), co-owns the media-holding *Kommersant* and the TV channel 7TV. ONEXIM Group (a private investment fund with interests in energy, mining and other industries) owns the news agency RosBiznesKonsalting, the newspaper *RBK Daily* and the magazines *RBK*, *M2*, *Nashi Den'gi*, *Autonews* and *Lifetime*. An even bigger company, but with similar interests, Interros, owns Prof-Media, which manages the TV channels TV3, MTV-Russia and 2x2, the radio stations *Avtoradio*, *Energy*, *Radio Romantika* and *Humour FM*, the newspaper *Afisha*, the Russian search engine www.rambler.ru and the websites www.lenta. ru, www.afisha.ru and www.101.ru (Media Atlas website 2011).

The climate change programme coordinator in the Russian branch of an international NGO states that these specifications of the Russian media ownership structure were the main reasons why so often (until a few years ago) newspapers and TV would report that climate change was a lie, a deception created by the West (interview 27, Moscow, 22 July 2011). The media outlets with an ownership structure significantly dominated by the state are obviously much more heavily influenced by the national climate policy. For instance, the predominantly state-owned information agency RIA Novosti (which will be discussed later on in greater detail) started to cover the climate change topic shortly after President Medvedev introduced the Climate Doctrine and made his appearance at the Copenhagen Conference in 2009. As a journalist of another state-owned media outlet said:

> I was almost forced to write about environmental problems around three years ago. My editor told me to do it, which has never happened before. I guess there was some kind of task set for my management. When we started our project [series of video news on environmental problems], even people in the city council said that probably 'federals' [state officials at the federal level] started to care about it [the environment].
>
> (interview 21, Chemal, 13 August 2011)

Nevertheless, in the situation of strong interdependence between the energy sector and the state, it always has to be considered with caution whether it matters if the media outlet is straightforwardly owned by the government or by, for instance, gas giants like 'Gazprom'.

Concluding, in Russia the majority of the media belongs to people or organisations whose main interests are outside of the media industry, and since media entities themselves are largely unprofitable businesses, their owners 'see media first of all as weapons to gain political capital' (Koltsova 2001, p. 322). But it is very difficult in Russia to separate the state from business, so regardless of whether the newspaper is owned by the government directly or by a large corporation, the goals of its owners might be the same. Lastly, as was demonstrated by examples in the case of climate change coverage, the ownership of Russian media is also influenced by the fact of the merging of the state and the energy sector – the main contributor to Russia's GHG emissions.

Advertising

In Russia advertising came into the media sector relatively recently, along with the collapse of the old state system. The old ideology of production, which was supported by the state and should have benefited its prosperity, was replaced by the new ideology of consumerism, which aims to satisfy individual needs (Grabel'nikov 2001). This thought is echoed by Oates and McCormack (2010, p. 122), who claim that 'there are two significant trends in Russian media content, one linked to market forces and the other to political pressures'. Political pressures have been discussed above in the 'Ownership structure' section and will be discussed further on in the chapter, whilst 'market forces' are the most relevant for this section in the context of the 'advertising barrier'.

Due to the emerging ideology of the free market, Russian media have been adjusted accordingly. For instance, politically oriented newspapers lost their circulation and can exist only by relying on the money of their supporting party. The so-called informative-commercial press has the biggest circulation and popularity. This type of Russian media organ is very close in its format to the Western tabloids, which are eager to attract their readers by writing about scandals, show business or sensationalist stories (including climate change). The informative-commercial press is very flexible and depends on the market's supply-and-demand logic, and that is why its quantity is always changing. Some papers replace others or change their content to fit audience interests (Grabel'nikov 2001). Ironically, these media organs are considered to be independent because they can exist without state support by giving lots of space to advertising. However, by gaining their independence from state ideology, the media are becoming the most powerful tool for the propaganda of the ideology of consumerism. This said, it should be acknowledged that when advertising was first introduced in Russia in the early 1990s, it took some time for people to get used to it and for industry to adopt it, so for the majority of media organs, advertising remained just a supplementary source of income.[6]

Overall, the advertising market in Russia has steadily grown over the last decade. Krylov and Zuenkova's (2003) analysis has showed that by 2003 advertising growth exceeded that of the country's gross domestic product by five times. Even though advertising budgets still might be lower than in countries where advertising is very important, like the United States, the market is increasing and cannot be ignored. Viktor Kolomiets, a professor in the faculty of journalism of the Lomonosov Moscow State University, explains the growth of advertising in Russia by such factors as political stabilisation of the society, which led to the growth of investments, including in advertising, the growth of the purchasing capabilities of the audience and an increase in competitiveness among businesses (see in Krylov and Zuenkova 2003).

Grabel'nikov (2006) vigorously argues that throughout the years of market reforms, Russian media also acquired the market features of 'bourgeois journalism' by copying the Western models, where ratings and advertisers take the first place and move the audience's interests to the background. From another point of view, a Russian 'special way' of doing things could be observed in this case as well, as Russian media did not simply copy the Western way of conducting media business, but reproduced it adjusting for its own national context. For instance, in the Russian case, the boundaries between the powers of advertisers, owners, sponsors and investors are blurred, but owners still remain at the top of this hierarchy.

A number of scholars (see e.g., Edwards and Cromwell 2006, Hearns-Branaman 2009, Lewis 2010) have noticed that advertising substantially influences the coverage of environmental topics in general and climate change in particular. For the Russian media, this does not seem to be the case. For example, a TV news journalist shared her experience on writing an article which created tension with the channel's advertisers. Some time ago she found out that one of the industries not far from her city had major faults in utilisation of its waste, hence she decided to raise the alarm, but the channel's bosses asked her 'to be nice' to the company, because they had a good contract with this organisation – 'we help them with PR and they pay'. Despite the 'recommendation', she still prepared her programme as originally planned, after which they indeed had problems with this company, but it did not last long and their relations were soon reestablished and carried on (interview 21, Chemal, 13 August 2011).

Other interviewees also denied noticing advertisers' direct influence on their everyday work. As the former news editor at a national TV channel suggested, 'I doubt that advertisers are only interested in numbers [ratings], but they want to know a bit about the content – for example, they would avoid programmes with open criticism of the Kremlin. Still, disagreements with the advertisers will not cause much trouble; the media outlet might just lose a bit of money' (interview 2, Moscow, 22 July 2011). Perhaps barriers created by advertisers are more of an issue for the editors than for the correspondents, or in the Russian case the influence of the latter is quite marginal in comparison to other factors.[7] Belin (2001, p. 328) argues that the number of media outlets in Russia (in particular in Moscow) does not represent 'the consumer demand or the size of the advertising market', because of the priority of the political use of the newspapers, where owners are willing to take financial losses in order to have access to the 'political' mass media (see more in Vartanova 2012).

Another specific characteristic of the Russian filter is that the advertising market is quite centralised and connected with ratings, mostly in the central part of the country or large cities (for example, Moscow or Novosibirsk); on the outskirts it is more hectic and is missing any kind of systematisation (Koltsova 2001). The regional media outlets struggle to maintain contact with sufficient amounts of advertisers, and to gain their loyalty media become much more dependent on advertisers; as Koltsova (2001, p. 324) concludes, from her research based on extensive fieldwork in Russia, this situation leads to a 'large amount of hidden advertisement'. This way of solving financial problems was especially popular during the early 1990s by writing articles with hidden messages promoting certain interests within them (so-called *dzhinsa* or *zakaz*) (Zhukova 2007). When journalists and editors did so, they were getting extra profit (Belin 2002). These articles were paid for unofficially – directly to the editor or journalist – and unfortunately sometimes they contained some aggressive accusations against certain political figures[8] or businessmen. Since they were disguised as objective journalistic opinion pieces, they would gain greater audience attention than the official advertisements. Currently this problem still exists to some extent, though the quality of articles has significantly improved (Koltsova 2006).

In conclusion, the advertising industry in the Russian media system has some very specific features. These start with the very recent appearance of advertising as an industry in Russia and its slow development due to the economic problems the country faced in the 1990s, and end with the blurred boundaries between advertisers' and owners' interests and even more with the merging of the state and the advertising industry[9] (Koltsova 2006). This last point fits within the larger problem of the extremely close interrelations between the state and business, which will be discussed further on in this book with regard to the energy sector. It can be suggested that in the Russian case, the advertising influence on its own is quite weak (which was also confirmed by the interviewees) and can quite easily be ignored in some situations.

Sources of information

Under Soviet rule, journalists were immensely restricted in the ways they could find information on the topic or receive comments from the parties involved, which led to the situation where 'a limited flow of information was the norm' (De Smaele 2007, p. 1300). Due to the journalists' inability to fully inform people of the acute problems, the existing informational gap filled up through informal ways of communication, as Soviet people through their personal connections tried to make sense of current political or social events.

After 1990, citizens' right to seek and obtain information (as well as distribute it) was recorded in the new Russian Constitution (1993), as well as in the Law on Mass Media (1991). Theoretically from that time onwards, journalists can 'knock on any door' and ask almost any questions they wish and nobody should prevent them from doing so or refuse to give information, as this would be considered illegal (of course, apart from some sensitive topics such as personal data and state or military secrets, which are specified in the law[10]). Furthermore, the change in the Russian media system at the end of the last century was accompanied by the establishment and development of a new professional activity – public relations. Russia as well as Western countries encountered the same problems here: various 'press centres, press services, press secretaries [. . .] were intended to facilitate journalists' access to information' (De Smaele 2007, p. 1301). In reality PR services became another obstacle in the way of obtaining data, since their goal was to provide information which only or largely benefited their clients. Governmental, commercial and other press centres tend to face opposite problems to those of journalists. They compete with each other to be able to get their information published or broadcast – it is not a secret that to present a company's information as news is a free and more effective way to advertise the company. At the same time, only certain information needs to be popularised, whilst other information needs to be hidden or at least presented in the most beneficial way (Chumikov 2001).

The phenomenon of journalism being replaced by PR technologies in Russia has often been criticised by scholars studying the Russian mass media. Skilfully prepared information by corporate media specialists does not leave much space for the journalists' investigation, analysis, or reflection on the problem (Bogdanov cited in Grabel'nikov 2006). Arguably, it leads to the degradation of the profession as such and allows sources of information not only to provide valuable opinions on the subject matter, but also to dictate the way it will be written and delivered to the audience.

De Smaele (2007) argues that one of the specific characteristics of journalism in Russia is the relatively important role of gaining information through personal connections. She refers to a study conducted in Voronezh in 2002 on the subject of the usefulness of personal connections for journalists. The results of this research showed that 70 percent of official written inquiries from journalists and the public were declined, whereas only 36 percent of those inquiries which were made through personal approach got rejected. Interestingly, as Konovalov (see De Smaele 2007, p. 1303) suggests, in the Russian case informal communication is the way to work within the constraints of secrecy mentioned earlier (such as vague definitions of state, military and business secrets). Furthermore, Russian journalists tend to have some kind of a rank attached to each newsmaker. Katja Koikkalainen (2008) suggests that for journalists, the most preferable sources would be the ones with the highest position in the organisation or with whom journalists have informal connections.

Another peculiarity of the Russian mass media is that they are concentrated in large media holdings and are controlled by a few owners, hence media outlets are united horizontally (various newspapers share the same owner) or vertically (different types of media such as TV channels, radio and newspapers are part of one media entity). As a result, different mass media organs receive information from the same source (Yushchenko 2007) – for instance, from the same information agency, such as RIA Novosti, which will be discussed below as one of the major information sources on climate change in Russia.

Informing about climate

Considering the scientific complexity of the climate change topic, the choice of information sources is extremely important and at the same time challenging. If journalists do not have (easy) access to climatologists who are willing to explain all perplexities of the environmental problem, they are forced to find other (often less reliable) sources (Boykoff 2011, 2013). Hence, in Russia (as in any other country), one of the main information sources is the 'expert' – climatologists or other natural scientists who directly study the problem on a daily basis and supposedly have the most up-to-date and objective information. Representatives of environmental NGOs such as WWF-Russia, Greenpeace Russia, Oxfam-Russia and so on can also act as experts.

Throughout the series of interviews with journalists conducted for this project, several conclusions have been reached with regard to the information sources for the climate change topic. Firstly, Russian climatologists are not very public people and journalists find it quite difficult to get in touch with them outside of the times they meet scientists during climate change conferences. The main scientific sources of information on climate change in Russia are the Voeikov Main Geophysical Observatory, the Federal Service for Hydrometeorology and Environmental Monitoring (Roshydromet) and the Institute of Global Climate and Environment, whose members often become part of the Russian delegation at UN conferences on climate change.

At the regional level, the situation is much worse – journalists struggle to identify whom they need to approach. A correspondent of a regional newspaper admitted that when she started to write about climate change, she could not find any experts in her city and she was forced to look for sources in other cities and regions (interview 8, Chemal, 13 August 2011). Sometimes journalists cannot get the necessary information because scientists do not want to give their opinion if it goes against official interests (various interviews, July-August 2012). As a result, journalists often remain alone with their problem: 'Even if we [journalists] understand that officials or businessmen do something wrong like damage the environment, we cannot object to them because our opinion is not qualified on the topic' (TV news journalist, interview 21, Chemal, 13 August 2011). Another problem of addressing scientists and experts as information sources was voiced by the correspondent of the Russian branch of an international broadcaster (interview 5, Moscow, 25 July 2011), who stated that one of the main reasons behind the low coverage of climate change in Russia is a disagreement within the scientific community and even a great degree of scepticism among Russian scientists, which was quite popular until very recently (for more on this problem, see Chapter 3 on Russian state policy).

In the case of climate change coverage, NGOs play a great role – for instance, the prominent climate change spokesperson and climate change programme coordinator at the WWF-Russia, Alexey Kokorin, has become one of the most quoted people on climate change in Russia. His expertise on the subject and his skills in communicating with the mass media help him to build long-term and mutually favourable relations with journalists. Kokorin states that his motto in communicating with media is 'never say no to them', so he always tries to explain and advise journalists on the problem, which will hopefully contribute to better coverage. However, he adds that this rule does not apply to TV talk-shows or scandalous cases where journalists try to create a scary spoof story rather than discuss the real problem (interview, Moscow, 27 July 2011). Caution when providing information for TV journalists was also raised by another NGO representative, who said that it is easier for him to deliver the message to print media, where he has more control over the final outcome, whilst in TV programmes words often get taken out of context and do not fit within the overall content of the programme (interview 19, Moscow, 27 July 2007).

Interestingly enough, in some cases during the major international conferences on climate change when the official Russian delegation fails to provide any kind of information for mass media through NGO representatives, Russian journalists themselves help to deliver the Russian official position on climate change to journalists from other countries. This paradoxical situation is a result of the poor publicity of the official Russian delegation at the UN conferences. As a news agency correspondent noticed, 'it seems that the Russian delegation has a position that it is better not to say anything at all in order to not get unwanted questions' (interview 30, Moscow, 22 July 2011). Another criticism of the Russian official delegation as an information source relates to the composition of the delegation: 'during the conference on climate change in Bonn, the American delegation, which did not sign the Kyoto Protocol, had 25 members whilst the Russian one had only 7, and they simply did not have expertise on some questions' (environmental NGO representative, interview 27, Moscow, 22 July 2011). In some cases, NGO activists even try to defend Russian officials. As one of them explained this paradox: 'when we are here, in Moscow, I always criticise our officials, but in front of media from other countries, it becomes more important to explain Russia's position rather than let it be blamed for all the sins' (interview 31, Moscow, 27 July 2011).

In general, Russian environmental NGOs 'use' international conferences to attract as much attention to the problem as possible by creating special webpages, blogs, press releases and other print or electronic material. It seems that only in exceptional situations do Russian NGOs find themselves in demand by journalists, whilst during 'quiet' times, attracting their attention becomes a struggle: 'especially until the end of 2009, almost nobody was interested in the topic despite all of our 'inventions'. We tried to organise various events and action days, but the results were almost negative' (environmental NGO representative, interview 27, Moscow, 22 July 2011).

The above-mentioned organisations are mostly located in the central part of Russia (mostly Moscow); in other regions, the situation is even less optimistic. The attention of mass media is even more difficult to attract, especially because it is more difficult to make the topic relevant to specific geographical areas. However, NGOs have found another way to provide information on climate change and attract attention to the problem by organising seminars devoted to climate change problems for journalists. In August 2011 the NGO 'Centre of Environmental Innovations' (Tsentr Ekologicheskikh Innovatsiy), with the financial help of USAID and the local support of the NGO Altai Regional Public Fund 'Altai – 21st Century', organised a media-training event: 'Les i Izmeneniya Klimata: Problemy i Resheniya' [Forest and Climate Change: Problems and Solutions] (which this author attended). The training took place in the picturesque village of Chemal in the Republic of Altai, where over four days journalists from the various central and regional media outlets learned how to cover climate change-related topics and had a chance to talk to experts and get information firsthand on climate change's consequences for Russia.

According to one of the organisers of the seminar, Elena Surovikina, their goal was to bring together journalists from different regions (centre and periphery) and different types of media (print, TV, radio), so in addition to other outcomes of the seminars, they would share with each other their own experiences of working on this topic (interview, Chemal, 14 August 2011). By the end of the seminar, each journalist had prepared at least one article devoted to climate change, but for the organisers of the training, the main goal was to educate journalists about the problem and interest them in the long run, so that they would pay greater attention to the problem throughout their careers. They are convinced that the seminars they had organised before significantly improved not only journalists' knowledge and understanding of the problem but also the general level of awareness of the problem among the population (Andrey Stetsenko, Centre of Environmental Innovations, interview, Chemal, 14 August 2011). Journalist-participants also consider this kind of activity a way to solve the problem of scarce sources of information on climate change, since it allows journalists to understand what the problem is about and whom they should approach in case they want to cover it, and what the consequences of climate change are for everyone, including people who read their articles or watch their programmes (various interviews and observations, Chemal, 12–15 August 2011).

State official information sources – main newsmakers on climate change?

'Impenetrable!' ('Neprobivaemye!') – this was how a journalist from a regional radio station described the local officials in their role as information sources (interview 17, Chemal, 13 August 2011). She added that when she tries to talk to officials about such sensitive topics as environmental change, it is extremely difficult to get through to them:

> In our city, there is a divide: representatives of the city authorities are very closed for communication (they demand endless confirmations with their press service), whilst regional authorities easily agree to give a comment, but their usual response is that everything is good and that they are working on all problems, so it is almost impossible to get the real information out of them.

The journalist admitted that eventually she stopped trying to organise meetings with them, since it always comes down to the official line rather than any kind of discussion. The climate change host at the radio station 'Voice of Russia', John Harrison, shared a similar opinion:

> It is very difficult to find somebody in the government to take part in my programme. Because until recently [Medvedev's announcement of the anthropogenic character of climate change], half of the government did not consider

that climate change would be a problem at all, therefore appearing on the programme on climate change would be counterproductive to them, and the other half are afraid to take part in any media show or articles.

(Skype interview, 18 June 2012)

From another side, quite often interviewees stated that in one way or another the main 'newsmakers' in the country for the topic of climate change were President Dmitry Medvedev and Prime Minister Vladimir Putin (their posts at the time of the interviews). There is a mutual agreement that in the last few years the climate change topic has caught on mostly due to the 'right' information coming from the heads of state: 'it is already good news for us that Medvedev started to admit the existence of the problem – the Conference was somehow covered only because at the last moment Medvedev decided to go there' (the energy efficiency project campaigner at a leading environmental NGO, interview 19, Moscow, 27 July 2011); 'the situation changed after the Copenhagen Conference, Russia saw economic benefits and the President announced a new direction' (correspondent for a national newspaper, interview 7, 7 July 2011); 'I would like to believe that our main newsmaker [Putin] will change the situation and make climate change a public problem' (correspondent, a national news agency, interview 31, Moscow, 20 July 2011).

All of the above-mentioned interviewees and a few others also expressed hope that more attention would be paid to the climate change topic with the appointment of the president's adviser on climate change issues, Professor Alexander Bedritsky (see Chapter 3 for more about this appointment). Because of the importance of the Russian official elite in the question of climate change coverage, one of the main concerns which arises out of this situation is whether the newly elected President Putin will continue Medvedev's more educated and more coherent policy and stance on climate change and will keep Bedritsky as his adviser or let us bear witness to more comments such as: 'wind turbines kill worms' and 'less money will be spent on fur coats'.

The role of information agencies as information sources – case study of RIA Novosti

According to Koltsova's (2006) research on the Russian media system, the hierarchy of the information sources is headed by the information agencies. Indeed, information agencies very often become the starting point for journalists writing articles or conducting independent investigations. The Russian international news agency RIA Novosti is one of the biggest agencies in the country; it also became one of the first media outlets to devote a separate subsection to climate change problems ('Pogoda i klimat' – 'Weather and climate')[11] within its bigger section on the environment. RIA Novosti has been providing news for over 70 years and started as the Soviet Information Bureau in 1941 with the main purpose of delivering news from the battlefields of the Second World War. Currently, it provides information for Russian and foreign mass media, the presidential administration,

Russian central and regional governments, various ministries and diplomatic services and NGOs, as well as numerous business organisations (RIA Novosti 2012).

Through its website, RIA Novosti targets ordinary people who prefer to look for their news on the Internet. In spite of RIA Novosti's long history and an impressive record of service, the climate section was launched quite recently, in 2009. On a weekly basis the agency usually publishes one or two information articles on climate change or a topic related to it, excluding times when something extraordinary happens such as international conferences on climate change, in which case the number of news articles might rise significantly. All news could be divided into several thematic blocks: Russia's involvement in the problem (acceptance and realisation of the Climate Doctrine, the work of Roshydromet, Joint Implementation projects), international negotiations, scientific findings, reports produced by the United Nations Environment Programme or IPCC and so on.

According to RIA Novosti's special climate correspondent Olga Dobrovidova, since these news items are very narrow and quite complicated, only specific people are involved in covering them (interview, Moscow, 20 July 2011). Thus around four years ago (when the section on climate was opened), Dobrovidova was appointed to work specifically on this topic, which has made her one of the first (and very few) journalists in Russia who specialise in topics related to climate change. The extremely limited number of Russian journalists writing about this topic is shown by their representation (or underrepresentation) at the UN conferences on climate change, where occasionally Dobrovidova meets a couple of her countrymen but mostly gets 'attacked' by her colleagues from other countries who are genuinely surprised to see a Russian journalist and very curious to hear how this topic is covered in Russia. The initiative to create such an unusual position for Russia as a 'correspondent-climatologist' was triggered by the UN Climate Change Conference in Copenhagen in 2009 (COP-15), as Dobrovidova said:

> before it nothing was practically written about climate change (in Russia), but after the Copenhagen [Conference] it became apparent that from now on it will be discussed a lot. Before nothing was happening in Russia, but after the Climate Doctrine's acceptance, it became a topic[. . . .] However, I am not sure if there was an 'order' [by the state] on the topic per se.
>
> (interview, Moscow, 20 July 2011)

Dobrovidova describes one of the main challenges in writing about climate change as the 'high entry barrier' – the topic demands an understanding of quite sophisticated issues and, in the end, without a degree in natural sciences, 'you just have to believe that there is a consensus that climate change is happening' (ibid). Even though RIA Novosti acts as an information source for many other mass media entities, finding its own sources on climate change issues becomes a problem for the agency. Firstly, it is quite difficult to get a press-release from the relevant scientific institutions or organizations with sufficient expertise on climate change and it often takes lots of time to look for the necessary people. Also, it is difficult to provide a balanced picture, since climate sceptics in Russia are 'insane' and they mostly think that climate change is a plot against Russia (ibid).[12] 'So in the end we just translate the official flow of information, trying to add some information from abroad, but we cannot create a proper discussion on the topic, though that is not our job to do. We are just an information agency' (ibid). Perhaps at the moment RIA Novosti is one of the most influential media outlets in terms of covering climate change problems. However, since the agency is also partly owned by the government, its impartiality is often questioned. Thus one of the interviewed environmental activists admitted that her world-famous organisation is struggling to get the attention of RIA Novosti, as she claims: 'they are used to working with the WWF-Russia and Greenpeace Russia, but they are not sure if we are not "harmful" [for the state]' (interview 27, Moscow, 22 July 2011).

In conclusion, as much as the 'sourcing' filter has some internationally common features due to the nature of journalism and the growing institution of PR, when analysing media coverage of climate change in Russia, it is important to keep in mind the specifications of the Russian media. De Smaele (2007) argues that due to the restricted access of information during the Soviet period, information became an 'elitists' commodity whose flow was controlled by a 'powerful minority' (ibid, p. 1310; see also De Smaele 2002). In the modern era, the situation has drastically changed; however, the selectivity of the information flow and privileged access to it remains. Even in the case of global problems like climate change, which affects absolutely everyone regardless of social status and level of knowledge on the subject, Russian journalists struggle to find information. Experts possess valuable knowledge but prefer not to get involved in such politicised issues. News agencies are very useful, but most of the time they are good only for journalists to get a general idea (it is a starting point for the further development of the story). Russian NGOs are not so authoritative (especially on the regional level and especially in comparison to their foreign colleagues), so the media often discount them or are reluctant to work with them. In the end, the most authoritative newsmaker becomes the state, which to a great degree (on purpose or not) manages the information flow on climate change.

Flak

In the USSR, 'flak' (mostly in the form of censorship) was a well-established practice. Methods of Party control could have included editors' regular meetings and reports to the Party representatives or, on the other hand, presentations by the Party's secretaries at the editorial meeting with clarification of what should be featured in the new issues.[13] During the *perestroika* era, these methods started to lose their effectiveness (from the point of view of CPSU leadership); and in the cases where they did work, they slowed down the democratic processes that had appeared in the media. Specifically, this was apparent when media tried to follow the principle of a plurality of opinions. Local Party committees still preferred to see on news pages the ideas which would not discredit the Party's leaders. At times they would even destroy all issues of the newspapers which they found outrageous. Voltmer (2000, p. 472) points out that 'the press under Gorbachev was still controlled, albeit the style of supervision changed from a confrontational to a cooperative relationship'.[14] Grabel'nikov (2001) argues that these control methods, in the end, played against the Party. The more media were forced to be a propaganda tool in support of the state, the more this backfired against the government. The media became one of the first institutions which turned against the Party and criticised it with even more power.

As *perestroika* progressed, the media gained more and more freedom and, arguably, a few years after 1989 enjoyed a 'honeymoon' (Voltmer 2000, p. 472) period of press freedom in Russia.[15] In 1991 for the first time in history, freedom of speech and expression was legally defined in the form of the state law on mass media, which 'prohibited censorship and barred government from shutting down media outlets [. . .] except by court order' (Coyne and Leeson 2009, p. 10). Unfortunately, this fundamental change did not bring the expected freedom, and despite legal restrictions, it is commonly accepted that although during the hectic years of the 1990s Russian media experienced some degree of freedom, eventually for different reasons it kept losing its autonomy. Currently many agree that censorship does exist in Russia, and it can be more or less obvious depending on the importance of the covered topic, whose interests are involved in it and the significance of the media outlet.

Dewhirst (2002) quotes six types of censorship in Russia (which were first discussed in print in 1996 by Russian scholar Aleksei Simonov). They are administrative and economic censorship (the officials' power to control resources needed for media operation such as printing plants and their influence on businesses to advertise or not to advertise their products in the media), censorship resulting from actions by or threats from criminals (there were various cases of murders of Russian journalists which arguably were connected with criminal showdowns),[16] censorship resulting from editorial policy and editorial taste (which might range from how an article fits within the media outlet's overall information policy to personal preferences of managerial boards which then get imposed on journalists)[17] and, finally, self-censorship.

Oates (2007, p. 1288) states, 'Russian journalists have a finely developed sense of self-censorship and self-survival' and this awareness of their own limits derives from 'the Soviet experience of journalists in which the action of a censor was rarely needed, as Soviet journalists understood the party "line" and the way all stories should be formulated by the time they received their first job' (ibid, p. 1286). Supporting evidence of this can be found in an article written by the former Soviet journalist Somov, in which he confessed that a censor was 'planted' by the Party inside everyone's soul, and this inner censor was worse than the official censor from outside. He explains it as follows: you could have tried to argue against censorship from outside, but nothing could have been done when you sincerely believed in the necessity of the rules imposed by the system (see in Strovskiy 2011). As a result, 'according to a survey conducted by [the Russian Union of Journalists] in 2005–2006, more than 80 percent of Russian journalists [. . .] faced different forms of censorship in their everyday work, and almost all admitted to self-censorship' (Azhgikhina 2007, p. 1259, citing Yakovenko 2006). The chief editor of *Ekho Moskvy*, Alexei Venediktov, stated that with regard to self-censorship, 'the key taboo topics are corruption among the elite and Chechnya, particularly the abuses by the Russian troops and pro-Moscow Chechens' (see in Orttung 2006). Further on, by referring to the data collected throughout the series of interviews, it will be argued that any open or direct forms of censorship are not quite relevant for the media coverage of climate change; however, self-censorship may still be important.

Censoring climate

During interviews with journalists from various media outlets and different geographical locations, interviewees often stated that in their work of covering environmental problems in general and climate change in particular, they do not experience any kind of censorship. 'Climate change is such an abstract topic, I cannot imagine a situation where an editor would tell me not to write about it', – says a TV journalist (interview 21, Chemal, 13 August 2011). She admitted, however, that she had extremely negative experiences with covering another environmental topic, when she was threatened by the managers of the organisation she was writing about. The journalist also noted that there are some 'political issues' which are implemented in the editorial policy, but journalists (especially ones who write about the environment) do not really notice it. For instance, if 'tomorrow the governor wants to come to our studio and talk about his work, he will be able to do it; of course, money is involved, but it does not concern us [journalists]' (ibid). Another journalist said that it is difficult for her to say if there is any censorship and that she always tries to discuss problems as objectively as possible, although she did admit the possibility of censorship amongst her colleagues (interview 30, Moscow, 20 July 2011). The situation of censorship on writing about environmental problems was well summarised by a correspondent of the regional newspaper, who insisted that nobody hovers over her, nobody will tell her what and how to write: 'they trust me and think that I am more knowledgeable about this subject'; however, she also admitted that if she goes too far and topics intersect the interests of the big industries, military forces or government, she finds herself alone in a confrontational situation:

> Once I wrote an article about waste management problems in the city and my editor was very happy with me; he even took my article to a presentation at an international symposium. But when people involved in this problem started to threaten me and filed an action against me, my editors said to accept the charges and pay the fine even though I was right, and of course the fine would come out of my pocket; the newspaper would at best pay half of it. So, I feel that I am free in my actions, but in a situation like this I find myself one-on-one with the problem.
>
> (interview 8, Chemal, 13 August 2011)

Of course the example given by the journalist might be more relevant for regional media outlets, as one newspaper correspondent confirmed – the farther away from the centre, the more journalists are restricted in their work (interview 7, 7 July 2011) (see also Belin 2001).

The Internet – a road to freedom?

In Russia just a few years ago, high expectations had been assigned to this new form of electronic communication (see e.g., Vinogradov 2006). As Yushchenko (2007) concluded, even though in modern Russia the political and financial elites dictate the rules of the media production process, the Internet might serve as an alternative. Its interactivity, lack of censorship and possibilities for open discussions attract a broad audience. As time has gone by, it has become more apparent that the Internet's role and degree of freedom were exaggerated and it did not really become a 'saviour' of Russian freedom of speech.

It is quite obvious that all of the types of censorship mentioned earlier to some extent contribute to Russia being placed by the Press Freedom Index (2011–2012) in the 142nd position out of 179. In 2012 the reasons behind such a negative situation were explained by the state's influence over media information and journalists' inability to freely perform their work, with journalists in extreme situations of struggle for freedom of speech, such as the numerous cases of reporters getting killed whilst their murderers were not always punished.[18]Among other reasons, the lack of diversity in TV and radio news was named by the Press Freedom Index, but recently even the Internet has become more and more restricted (Reporters Without Borders 2012).

In 2012 an article describing press freedom in Russia paid close attention to the censorship which now has spread to the Internet, which arguably was provoked by state officials' realisation of the Internet's growing significance. For instance, it is widely accepted that massive civil protests against the results of the Duma and presidential elections in December 2011 and March 2012, respectively, were to a great degree organised through various social networks and blogs. This in turn led to a negative reaction from the government, as a result of which many websites got banned and bloggers got sued.[19] Furthermore, sometimes online activity or the independence of online media is simulated by officials (anonymous source, interview 1, 27 May 2011). As Oates and McCormack (2010, p. 133) note, 'Russia is shaping the Internet, rather than Russian society being shaped by the Internet. This is a particularly clear and compelling image of how the Internet is constrained by domestic, rather than international, political communication norms'.

As far as media coverage of climate change goes, the Internet also does not play as great a role as could be imagined. Traditional media duplicate information on their webpages or blogs.[20] In turn, social media users republish the same information on their personal pages (Poberezhskaya 2014). This practice does not add new content to an already existing body of information on the subject matter, but at the same time, based on which material keeps reappearing on personal webpages, it does allow us to see the public's interest in certain aspects of climate change. For instance, as a brief analysis of Russian-language Twitter has shown (ibid), people are more likely to re-tweet information concerning the negative consequences of climate change which threaten humans' health or the population of some type of animals (for example, deer or snails). Social media also allow for expression of a more polarised opinion spectrum – starting with extreme sceptical opinions on climate change existence and ending with very proactive practical information on how we can stop the temperature rise. Lastly, the climate change mitigation discussions on social networks are usually started by the relevant NGOs and serve as another resource not only for expressing opinion on the topic, but also for journalists to get in touch with the necessary experts and to learn new information about the problem.

Furthermore, it should be noted that in Russia, environmental NGOs are often not powerful enough to be heard or noticed by the civil society or by the state in order to attract attention to certain problems (in Chapter 3, the role of environmental NGOs in Russia's climate change policy is discussed in more detail). The Internet allows them to express their reaction to some questionable media articles or programmes. Thus when in 2009 Channel One broadcast the documentary *The History of a Certain Lie, or Global Warming* [*Istoriya Odnogo Obmana, ili Global'noe Poteplenie*, directed by Sergey Nadezhdin], which debunked the 'myth' of climate change,[21] Greenpeace Russia instantly published online video responses and articles explaining how the documentary was misleading (Greenpeace 2009). WWF-Russia reacted the same way (Ecoloungetv 2009). According to Igor Podgorny from Greenpeace Russia, they also got in touch with journalists and with the producers of the TV channel with the appeal that such information should not be broadcast (interview, Moscow, July 2011). A year later a documentary with an absolutely different message on climate change was broadcast[22] and Russian NGOs took a big part in its production; however, it is extremely difficult even for NGOs themselves to judge whether it was solely their achievement or a result of the change in state policy.

In conclusion, scholars and practitioners agree on the presence of censorship in Russia, which indeed leads to a situation where in the weak civil society, 'flak', or reaction to media products, is produced by the state and even the development of the relatively new way of communication – the Internet – has not broken this pattern. It should be stated that the degree of censorship in any type of media depends on the importance and sensitivity of the topic. For instance, climate change, being so 'abstract' and until recently ignored by state policy, does not require much control. However, it is very difficult to prove or disprove whether the coverage of climate change has been influenced by journalists' self-censorship. In this sense, arguably it proves the existence of the 'subtle ways' of media control. Unlike with other sensitive topics, such as Chechnya and corruption, journalists write about climate change in a certain way not because they are forced or told to do so, but because working in a certain political and economic context, reporters willingly respond to the elites' interests.

Dominant ideology

Just over 20 years ago, the dominant ideology of Russia was very clearly defined and its constraints on the media system were acknowledged and even institutionalised. The understanding of the media as a powerful tool of propaganda was central for the government of the Soviet Union. The mass media were not part of socioeconomic relations, but middlemen which would accumulate ideas about the main doctrine and then pass them on to the people.

During the *perestroika* period, when the 'country was slowly moving from the totalitarian towards the fragmented political culture' (Strovskiy 2011, p. 235), the role of mass media changed and their everyday work routine was influenced by the surrounding ideological transformation. Further on, when the old state ideology ceased to exist, the media got an opportunity to become a member of society powerful enough to influence the processes happening in the state and to become a political institution on their own. Indeed, the media can be a way for people to express their view of the political situation. The media can also inform and educate people and, therefore, help them to build a democratic society.

The new ideology and, in particular, plurality in party representations in modern Russia made it possible for each political movement to be able to have its own media outlet and, hence, be able to state its ideas and programmes. However, these new political media did not have much power, they did not have enough *mass* to be noticed and to be able to make a change or convince people to be supportive of any particular political movement. Grabel'nikov (2001) also mentions the so-called hidden political affiliation of the media, which would not admit that they supported a certain side but would quite obviously deliver information in favour of that hidden owner or investor. *Nezavisimaya gazeta* (Independent newspaper) or NTV (Independent television channel) were independent only in name. In support of this, Grabel'nikov cites the words of NTV's former manager, Malashenko, who admitted that the word 'independent' in the abbreviation of NTV did not mean anything. The channel belonged at that time to the Russian oligarch Vladimir Gusinsky, who, according to Malashenko, had the right to exercise his power and fire the manager anytime he wanted to.[23] This situation gets worse depending on how far away regions are from the centre. During an interview, an anonymous source from Kemerovo's[24] city council who on a daily basis works with the media claimed that

> we do not have independent media; they all have their political agenda which is defined by the municipal or regional administrations. Journalists write according to our press releases, and their articles should not deviate from the 'party line'. Everyone knows about environmental problems in our region [high level of air pollution due to the coal mining industry] and how it damages our health, but nobody wants to touch this topic. It is out of the ideological frame.
>
> (interview 1, 27 May 2011)

At different stages of state development, the Russian media needed to find their place inside the state ideology, whether it was the implementation of the communist ideas, the popularisation of ideas of the free market and the ideology of consumerism and so on. The media never really became an independent power on its own – a 'fourth estate', where it does not fit into the ideology but starts to create it. This kind of power was only possessed by the Russian media for a relatively short period of time during the first Chechen war (Grabel'nikov 2001). In 1994 the conflict in Chechnya brought major disagreements in society where people took sides depending on their pro- or anti-war moods, and the mass media took a very strong anti-war stance and generated a strong campaign against the government and army (Zassoursky 2004). Grabel'nikov (2001) even reveals that in the war zone in Chechnya, the government was trying to intercept the signals of the radio stations because their messages had a negative influence on soldiers. The government lost the information war by failing to explain to the people inside the country and abroad what the purpose of the war was.[25] Even though for some time the media behaved as actors in domestic politics, quite soon the state understood its mistake. After this war, the media ownership structures were all reorganised again (see the previous section 'Ownership structures'). People in power realised that media could be just as powerful as they were, so they needed to be taken into account.

The current regime in Russia is often characterised as 'a managed democracy', in which all formal attributes of the democratic regime are in place (such as elections, a constitution, divided branches of power, plurality of political parties and active civil society); however, they do not properly perform due to the corruption and centralisation of power by the small group of elites (or even worse by one person – an extremely powerful president or prime minister, depending on the period). Richard Sakwa (2011) studies this duality of Russia's modern political regime and even defines Russia as 'a dual state' in which 'the legal-normative system based on constitutional order is challenged by shadowy arbitrary arrangements' (ibid, p. viii). For instance, when Putin throughout his two terms centralised and strengthened his presidential power, he did not break constitutional law and did not run for a third consecutive term, but instead found an obedient successor for his policies; and after allowing Medvedev to be elected president for one term, Putin again came into office in 2012 without officially breaking any laws.

The same example of duality can be used in describing the Russian media system. Maria Lipman (2009, p. 3) in her report on Russian media for Chatham House claims that 'in today's Russia [. . .] the media are reduced to being a political tool of the state or marginalized to a point of making no difference in policy-making'. She explains this by 'the lack of an enabling environment' or any place for 'political pluralism, the separation of powers and the rule of law' (ibid). Becker (2004, p. 149) also argues that a state-controlled media system like the one which can be observed in Russia is a sign of democratic degradation (the author defines this system as 'neo-authoritarian').

In this type of system, ownership of media is not restricted (as discussed above, Russian law allows anyone to own a media outlet); however, whilst state-owned media are quite openly controlled, media with other types of owners can be influenced through economic pressure or ambiguity in the law. Oates (2007, p. 1296) notes that due to the 'new controls and pressures on [Russian] journalists, notably market forces', the system of Russian media can be called 'neo-Soviet'. Whilst many changes have happened after the collapse of the USSR, the media remain 'a tool for the elites rather than a watchdog of the masses' (ibid, p. 1297). This idea is also echoed by Oates and McCormack (2010, p. 118), who state that neither society nor journalists or politicians see the Russian mass media as 'objective' or 'balanced'.

Nevertheless, despite all the criticism of the Russian modern mass media, Becker (2004) still stresses that the 'neo-authoritarian' media system should not be equated with a 'totalitarian' or even a 'post-totalitarian' one, since despite all of the restrictions and 'hiccups' discussed earlier, there is clear media variety, a legally supported media, journalists who enjoy independence and a new ideology which even now (with Putin in his third term) does not parallel the former communist regime. In the words of the former news editor of a national TV channel:

> Putin and Medvedev both support the atmosphere of freedom of speech, so you cannot deny its existence. If you want to criticise Putin, do it, but you need to support your statement. So it is all within the ideas of the 'law-based state', but then you might be called to the court to hold a response for your article and there because of the corruption and vague laws, you might pay for your words. So, freedom of speech exists under Putin and Medvedev – it is not a bloody regime, it is Pinochet with a human face.
>
> (interview 2, Moscow, 22 July 2011)

But the contradictions described do not just exist in the system as such but are even demonstrated throughout the statements of its main mastermind. Burrett (2011) provides an example of one of Putin's speeches on media given in 2000, where he states that 'without truly free media, Russian democracy will not survive' and at the same time shares his concern that by following their owners' (the oligarchs') interests, media become a 'means of struggling against the state' (ibid, p. 5).

As Strovskiy (2011) concludes, one of the reasons for restrictions on press freedom is that Russian media were unable to maintain their independence during the transformation phase from the ideology of socialism towards the ideology of the free market. In the course of two decades, most Russian media organs did not manage to find a way to achieve financial independence and became severely dependent on businesses, which in turn (in the 2000s) were taken over by the state. Once again we return to the specific characteristic of the current Russian political and economic system: the tight connection between the Russian state and business, where 'the state clearly rests on top of the food chain' (Becker 2004, p. 152). Soldner (2008) argues that these strong connections between state and economy in Russia are typical characteristics of 'political capitalism'. He borrows this concept from Max Weber, who makes a distinction between 'rational market-oriented capitalism' and 'politically oriented capitalism' (Soldner 2008 citing Weber 1980, p. 158). For example, in Russia, people go into politics to make profit in their businesses or control major industries in order to maintain political power. In this situation, the media are treated as a 'political resource' (ibid, p. 172) and their functions are to ensure communication with the electorate and promote the political interests of the interested parties. Soldner (2008, p. 160) argues that 'one of the most important consequences of "political capitalism" is that it suppresses the emergence and establishment of alternative societal actors, such as political parties, trade unions, independent mass media and NGOs', which is the exact problem discussed in the previous section on 'Flak'.

It is suggested here that in regard to the research of media coverage of climate change in Russia, the factor of the dominant ideology has to be studied in greater detail but in the context of the state's climate policy, which will be presented in the next chapter. The importance of the state has already been demonstrated in shaping other factors and actors of media coverage of climate change, and a detailed study of the climate policy will show whether media policy indeed was in consent with the state's position on the issues.

Concluding remarks

Starting with Lenin's appeal to use mass media as a 'propaganda tool', in order to mobilise the masses for the purpose of the revolution, the media in Soviet Russia indeed served as a 'tool' in the hands of the state leaders. During *perestroika* and especially the early 1990s, the media reached its 'golden age' (Belin 2002), when it even managed to influence some political and social processes, but as a prominent Russian journalist, Nadezhda Azhgikhina (2007, p. 1248) states, 'clearly, after the temptation of being the "fourth power", the media lost its real independence and quite quickly became a convenient tool for elite power and structure'. So, once again media are seen as and are referred to as a 'tool' (even by journalists themselves) and their freedom is questioned, but now there is no revolution to make and no communist state to build. Russia is supposedly an emerging democracy with, among other attributes legally established and guarded, freedom of speech. But despite the democratic reforms, as Azhgikhina (2007, p. 1246) pessimistically concludes, 'the media are becoming more and more primitive, combining propaganda and entertainment which is steadily edging out serious analysis, and free voices are hardly audible'.[26]

The key to this riddle of the controversy lies within the structure of the current media system. The state or large industries with close connection to the state own the majority of the influential media in Russia. In the case of climate change, this factor plays a crucial role, since the state's policy for a long time was very controversial and these industries come predominantly from the energy sector. The advertising market in Russia is not as significant as in its former (or present) rival – the United States. The key difference between the United States and Russia is that even though in both cases the media are dependent on financial investments from the business sector, in the United States these corporations are more independent from the state, whilst in Russia the line is blurred. As Bagdikyan (cited in Zhukova 2007, p. 42) states, the United States media have 'sacred cows' (owners, their family or friends, advertisers) which can influence any article, whilst in Russia these 'sacred cows' exist mostly in the form of the state authorities. With regard to the 'information sources', despite the fact that climate change is a scientific topic, state leaders were acknowledged to be amongst the main newsmakers in Russia. De jure censorship is banned in modern Russia, but de facto censorship exists in various forms. Even though the climate change topic is not significant enough for the state to censor it and, as Soldner (2008, p. 170) admits, 'where power is not at stake, the Russian mass media can and sometimes do offer a wide range of viewpoints', quite often journalists writing about various environmental topics face or choose to face 'self-censorship'. Finally, the new regime in Russia ('dual' or 'managed' democracy) puts the media within constraints where from one side they have to adjust to the new ideology of the free market, and from another side still have to coordinate to some extent with the authoritative power of the state.

As has been demonstrated in the case of media coverage of climate change in Russia, the process of 'manufacturing consent' does not need to be purposely controlled or forced, but because of the way the Russian media system operates, coverage would go up when Medvedev accepts the climate doctrine, when gas companies see economic benefits from climate change mitigation or when NGOs are not opposed by the government and are heard by the journalistic community. With the rising threat of global environmental problems such as climate change, media can be seen as a mechanism to stimulate or protect sustainable development. As Shumilina (2010) argues, eventually the media can lead to a change of paradigms of values and turn mass consciousness towards the new societal model (supposedly with the environment being more prioritised and being included into other spheres of life). I argue that change is happening, as various interviews demonstrated, but the question remains: what is the rationale behind it and how long will it last? In order to answer these questions, I propose looking at the changes and rationale behind the state's climate policy (which will be discussed in the next chapter) and the actual media coverage of climate change in Russia by analysing media texts devoted to this topic in Chapter 4.

Notes

1 Markov (2010) argues that censorship appeared in Russia long before the press was introduced – during the time of Alexis I (1645–1676). However, official censorship of the media appeared at the end of the reign of Catherine II (1762–1796), and under Paul I (1796–1801) it became formally institutionalised (see more in Esin and Zasurskiy 2003).
2 See more on the role of media during the period of *glasnost* in Mickiewicz (1999).
3 Becker (2004) states that due to Putin's policy, the Russian media situation significantly dropped down in the various ratings measuring freedom of speech. For example, in 2001 the Committee to Protect Journalists included him in the 'Ten Worst Enemies of the Press' list. Russia became one of only five countries on the list of the states 'endangered with becoming repressive' (International Press Institute's Watch) and Reporters Without Borders referred to Putin's media policy as 'too grotesque to be true' (ibid, p. 140).
4 Even though the owners of the major media entities often state that they try to stay out of media policy and their corporate arrangements, recent events which happened with the influential newspaper *Kommersant-Vlast'* tell the opposite story. After its provocative coverage of the parliamentary elections in December 2011 (with references to

frauds by the United Russia Party and a photograph of one of the ballots with insults towards Putin), its owner, oligarch Alisher Usmanov, recommended the dismissal of the editor-in-chief and the general director of the publication's holding company (Schwirtz 2011).

5 According to the former Russian Ministry of Press, in 2003 only around 10 percent of press media had relative economic independence (predominantly managed through the collective ownership of the journalists), whilst the majority of media organs belonged to the state or private owners (Nenashev 2010b).

6 Koltsova (2006) gives three reasons for the slow development of advertising in Russia: first, the low purchasing capacity of the audience; second, people were not ready for this new way of goods promotion and businesses did not have experience in how to use this tool to their advantage; and last, unfavourable legislation which was rather limiting for media trying to build their business on income from advertising.

7 It is interesting that even the limited influence of advertisers in Russia is perceived there much more negatively than that exercised by the media owners (Koltsova 2001). This may be related to the country's past, in which people were quite used to state control, whilst attributes of the new capitalist ideology are still new to them.

8 One of the most famous examples would be Sergey Dorenko's 'TV war campaign' against the then mayor of Moscow Yuriy Luzhkov and the Prime Minister Evgeniy Primakov, who according to different sources was one of the main opponents of Vladimir Putin (who was running for the presidential post). Dorenko's position was argued by a journalist from another TV channel – Evgeniy Kiselev. As Strovskiy (2011) states, this battle was motivated by political ambition of the TV channel's owners – oligarchs Berezovsky and Gusinsky (see more in Belin 2001, Zassoursky 2004).

9 Exemplary of the close connections between the state and advertising markets are the companies Premier SV and Video International, which throughout the last decade have had exclusive rights to sell advertising space on major national TV channels and have demonstrated close ties with the state (Belin 2001).

10 Interestingly, in the new law, ecological data have fallen into the category of information, which requires full exposure and cannot be hidden from the public, whilst during the Soviet Union, on the contrary, their dissemination was taboo (De Smaele 2007).

11 After the modernisation of the website in 2012, RIA Novosti does not have subsections within its main section, hence climate change news is now published together with other environmental news under the heading 'Ecology' (*Ekologiya*).

12 Interestingly, a few other journalists also evaluated the difficulty in finding an adequate sceptical position on the problem as a negative one. As was discussed in Chapter 1, by internalising the 'balance norm', which rather distorts the information on climate change, Russian journalists follow the trend which was observed in the Western media several years ago but which has been acknowledged as no longer relevant.

13 One of the most interesting artefacts of the Soviet era, whose government, of course, was capable of total censorship, was the *samizdat*, an off-the-grid circular through which freedom of speech was ensured: 'This involved the underground production and distribution of a wide range of media, including political and social commentary and full-length manuscripts on a variety of topics[. . . .] [T]he underground media provided alternative ideas to those the state disseminated through official media' (Coyne and Leeson 2009, p. 9). Quite often people involved in the production and distribution of *samizdats* were prosecuted; however, the existence of *samizdats* shows how people tried to find a way to express their opinions and fight the system, even if they did not succeed.

14 Former chairman of the Gosteleradio USSR (the state TV-radio network) and former editor of the newspaper *Sovetskaya Rossiya* Professor Mikhail Nenashev in his opinion peace for the newspaper *Pravda* in 1990 (Nenashev 2010a) wrote that Party members and leaders were not ready for such active criticism by media and advocated that critique (therefore freedom) should be limited and that it would be better if it actually came from the Party itself.

15 Mikhail Nenashev (2010b) states that now 20 years after the collapse of the USSR, it became obvious that the press was the main opposition and the most important tool in destroying the Soviet societal order. Ellen Mickiewicz (1999) also describes the media during *perestroika* as a 'central component' (ibid, p. 11); she refers to her interviews with Gorbachev and other members of his Politburo, where they all confirmed that 'every Politburo meeting started with the mass media' (ibid).

16 For a good discussion of the 'state and non-state agents of violence' in regard to the Russian mass media system, see Koltsova (2006).

17 Perhaps this type of censorship can be found in all countries around the world and thus, in some definitions, would not be 'censorship'.

18 See the report *Partial Justice: An Inquiry into the Deaths of Journalists in Russia, 1993–2009*, on murders of Russian journalists (International Federation of Journalists 2009).

19 One of the most notable scandals was raised around the infamous Russian activist and blogger Aleksey Navalny, who due to his online activity against corruption in Russia and more recently against the unfair parliamentary and presidential elections became 'an enemy of the state', which led to his arrest and imprisonment for 15 days and numerous cyber-attacks on his website (Ennis 2012).

20 The limited role of the Internet in the climate communication process is not unique for Russia. For example, Neil Gavin (2009, p. 130) in his research on the role of the Internet in UK climate change politics, states that 'for British citizens to make effective use of the web, they need to be a good deal more connected, interested, persistent, and "web-savvy" than they actually are. Consequently, its influence on climate change politics may still only be marginal.' The Internet does provide an open forum for various opinions and a vast amount of information on the topic; however, as Gavin (2010, p. 469) argues, 'the web perhaps generates more heat than light, its contribution to informed debate being mixed at best, and very unedifying, or even distasteful, at worst', and not many people have the skills and patience to work their way through the questionable or sceptical information on climate change. On the other side, often the Internet just re-duplicates the messages popularised by the conventional media (Gavin and Marshall 2011).

21 The documentary *The Story of a Certain Lie, or Global Warming* was broadcast on Channel One on 12 October 2009, two months before the UN Climate Change conference in Copenhagen. The provocative documentary was made on the basis of the material assembled two years beforehand by British television producer Martin Durkin for the movie *The Great Global Warming Swindle*. The movie aimed to assuage all worries that the Russian audience might have had. The leading storyline sought to demonstrate that climate change was a fiction created by political lobbies to promote their interests. The movie discussed Russia quite briefly at the end in the context that restrictions on GHG emissions were supposed to bring economic benefits to the country, which was going to trade its spare quotas; however, 'the promised millions were not received' (quote from the movie). Nevertheless, Russia was lucky that the USSR's economy had been at its peak of industrial development and the level of GHG emissions was enormous, then Russia's economy after 1990 experienced major decline, which allowed the country to meet the demands of the Kyoto Protocol without any problems.

22 The documentary *We'll Burn, We'll Freeze, We'll Survive* (Shilovskiy 2010) was broadcast on Channel One on 19 October 2010 as a reaction to the previous summer. As in the former example, climate change was the leading theme in the movie; however, the approach was different. The whole movie is full of alarming messages to attract audience attention to the problem of climate change by evoking their fears. Furthermore, the movie was created based predominantly on Russian data and explained very well Russia's vulnerability to climate change.

23 Currently NTV belongs to the Russian gas giant corporation Gazprom (Media Atlas website 2011). Incidentally, the politicians and businessmen of the new Russia very quickly came to the conclusion that if you owned television stations, then you had power. In the middle of the 1990s, when 80 percent of the population reached the poverty line, people did not have an opportunity to buy subscriptions to newspapers and magazines; therefore, radio and television became the only sources for free information. So, in spite of the freedom of speech coming into power, business and state elites brought almost all radio and television channels under their control (Grabel'nikov 2001).

24 Kemerovo is the capital of the Kemerovo Oblast, which is situated in the biggest coal mining area in Russia (the Kuznetsk Basin).

25 See more on the Chechnya coverage in Mickiewicz (1999).

26 Vartanova (2012), sharing a similar view, notices that 'entertainment became an attractive and politically risk-free content concept for many Russian media'. Journalist Grigorii Pasko, who himself was imprisoned for reporting on the environmental threat of nuclear waste and nuclear submarines of the Pacific Fleet, goes even further by referring to Anna Politkovskaya's posthumous article and the activity of Russian journalists as a 'farce of "clowns"', whose purpose was 'to entertain the public' and 'if they do write about serious matters, then they only say how great the "power vertical" is in all its manifestation' (Pasko 2006, p. 8).

Bibliography

Azhgikhina, N. (2007) 'The Struggle for Press Freedom in Russia: Reflections of a Russian Journalist', *Europe-Asia Studies*, 59/8: 1245–1262.

Becker, J. (2004) 'Lessons from Russia. A Neo-Authoritarian Media System', *European Journal of Communication*, 19/2: 139–163.

Belin, L. (2001) 'Political Bias and Self-Censorship in the Russian Media'. In: A. Brown (ed.) *Contemporary Russian Politics. A Reader*, Oxford: Oxford University Press: 323–342.

Belin, L. (2002) 'The Russian Media in the 1990s', *Journal of Communist Studies and Transition Politics*, 18/1: 139–160.

Boykoff, M. (2011) *Who Speaks for the Climate? Making Sense of Media Reporting on Climate Change*, Cambridge: Cambridge University Press.

Boykoff, M. (2013) 'Public Enemy No. 1?: Understanding Media Representations of Ecolounge/About the Film 'The History of a Certain Lie' Outlier Views on Climate Change', *American Behavioral Scientist*, 57/6: 796–817.

Burrett, T. (2011) *Television and Presidential Power in Putin's Russia*, Oxon: Routledge.

Chumikov, A. (2001) *Svyazi s Obshchestvennost'yu*, Moscow: Delo.

Constitution of the Russian Federation (1993), www.constitution.ru/en/10003000-01.htm, date accessed 2/4/2015.

Coyne, C. and Leeson, P. (2009) 'Media as a Mechanism of Institutional Change and Reinforcement', *Kyklos*, 6/1: 1–14.

De Smaele, H. (2002) '"In the Name of Democracy". The Paradox of Democracy and Press Freedom in Post-Communist Russia'; paper presented at the European Consortium for Political Research Workshop on Political Communication, the Mass Media, and the Consolidation of Democracy, Turin (22–27 March).

De Smaele, H. (2007) 'Mass Media and the Information Climate in Russia', *Europe-Asia Studies*, 59/8: 1299–1313.

Dewhirst, M. (2002) 'Censorship in Russia, 1991 and 2001', *Journal of Communist Studies and Transition Politics*, 18/1: 21–34.

Ecoloungetv (2009) 'O Fil'me "Istoriya Odnogo Obmana" Spetsial'no dlya Proekta' [video online], *Youtube*. www.youtube.com/watch?v=vyFZVhwiCs, date accessed 3/03/2014.

Edwards, D. and Cromwell, D. (2006) *Guardians of Power: The Myth of the Liberal Media*, London: Pluto Books.

Ennis, S. (2012) 'Profile: Russian Blogger Alexei Navalny', www.bbc.co.uk/news/world-europe-16057045, data accessed 6/03/2013.

Esin, B. and Zasurskiy, Y. (eds.) (2003) *Russkaya Zhurnalistika v Dokumentakh: Istoriya Nadzora*, Moscow: Aspekt Press.

Gavin, N. (2009) 'The Web and Climate Change Politics'. In: T. Boyce and J. Lewis (eds.) *Climate Change and the Media*, New York: Peter Lang Publishing: 129–142.

Gavin, N. (2010) 'Pressure Group Direct Action on Climate Change: The Role of the Media and the Web in Britain – a Case Study', *British Journal of Politics and International Relations*, 12: 459–475.

Gavin, N. and Marshall, T. (2011) 'Climate Change and International Protest at Copenhagen. Reflections on British Television and the Web'. In: S. Cottle and L. Lester (eds.) *Transnational Protests and the Media*, New York: Peter Lang: 197–209.

Gazprom-Media website (2013), www.gazprom-media.com, date accessed 20/02/2013.

Grabel'nikov, A. (2001) *Rabota Zhurnalista v Presse*, Moscow: Rip-holding.

Grabel'nikov, A. (2006) 'SMI Postsovetskoy Rossii: Pyatnadtsat' Let Spustya', *Vestnik Universiteta Rossiyskoy Akademii Obrazovaniya*, 1: 19–31.

Greenpeace (2009) 'Global'noe Naduvatel'stvo na Pervom', www.greenpeace.org/russia/ru/news/3972797, date accessed 4/11/2009.

Hearns-Branaman, J. (2009) 'A Political Economy of News Media in the People's Republic of China', *Westminster Papers in Communication and Culture*, 6/2: 119–143.

Herman, E. and Chomsky, N. (1994 [1988]) *Manufacturing Consent. The Political Economy of the Mass Media*, London: Vintage.

International Federation of Journalists (2009) 'Partial Justice: An Inquiry into the Deaths of Journalists in Russia, 1993–2009', www.ifex.org/russia/2009/06/23/ifj_partial_justice_report.pdf, date accessed 12/05/2012.

Koikkalainen, K. (2008) 'Journalistic Source Practices in Russian Business Dailies'. In: S. White (ed.) *Media, Culture and Society in Putin's Russia*, Basingstoke: Palgrave Macmillan.

Koltsova, O. (2001) 'News Production in Contemporary Russia: Practices of Power', *European Journal of Communication*, 16: 315–335.

Koltsova, O. (2006) *News Media and Power in Russia*, Oxon: Routledge.

Krylov, A. and Zuenkova, O. (2003) 'Reklama v Rossii: Sostoyanie i Perspektivy', www.dela.ru/articles/17419/, date accessed 25/04/2010.

Kuznetsov, I. V. (2003) *Istoriya Otechestvennoy Zhurnalistiki*, Moscow: Izdatel'stvo Nauka.

Law of the Russian Federation 'On Mass Media', 27.12.1991, N 2124-1.

Lenin, V. (1969) *Chto Delat'?* Moskva: Izdatel'stvo Politicheskoy Literatury.

Lewis, J. (2010) 'The Myth of Commercialism: Why a Market Approach to Broadcasting Does Not Work'. In: J. Klaehn (ed.) The *Political Economy of Media and Power*, New York: Peter Lang Publishing: 337–356.

Lipman, M. (2009) 'Media Manipulation and Political Control in Russia', *Chatham House Report*, www.chathamhouse.org/publications/papers/view/108964, date accessed 20/02/2012.

Lipman, M. and McFaul, M. (2001) 'Managed Democracy' in Russia: Putin and Press', *Harvard International Journal of Press/Politics*, 6/3: 116–127.

Markov, E.A. (2010) 'Vlast' i SMI v Rossii: Istoriya Vzaimodeystviya', *Izvestiya Rossiyskogo Gosudarstvennogo Pedagogicheskogo Universiteta im. Gertsena*, 122: 204–214.

Media Atlas website (2011), www.media-atlas.ru, date accessed 16/06/2011.

Mickiewicz, E. (1981) *Media and the Russian Public*, New York: Praeger Publishers.

Mickiewicz, E. (1999) *Changing Channels: Television and the Struggle for Power*, Durham, NC: Duke University Press.

Mickiewicz, E. (2008) *Television, Power, and the Public in Russia*, Cambridge: Cambridge University Press.

Nenashev, M. (2010a) *Illyuzii Svobody. Rossiyskie SMI v Epokhy Peremen (1985–2009)*, Moscow: Logos.

Nenashev, M. (2010b) 'Nezavisimost' SMI – Illyuzii i Real'nost'. In: S. V. Konovchenko (ed.) *Sovremennye SMI Rossii: Teoriya i Praktika. Sbornik Nauchnykh Statey*, Moscow: Moskovskiy Gosudarstvennyi Universitet Pechati: 5–18.

Oates, S. (2007) 'The Neo-Soviet Model of the Media', *Europe-Asia Studies*, 59/8: 1279–1297.

Oates, S. and McCormack, G. (2010) 'The Media and Political Communication'. In: S. White, R. Sakwa and H. Hale (eds.), *Developments in Russian Politics*, Basingstoke: Palgrave Macmillan: 118–134.

Orttung, R. (2006) 'Kremlin Systematically Shrinks Scope of Russian Media', *Russian Analytical Digest*, 9: 2–5.

Pasko, G. (2006) 'Black Mark for Journalists', *Russian Analytical Digest*, 9: 8–10.

Pasti, S. and Pietilainen, J. (2008) 'Journalists in the Russian Regions: How Different Generations View Their Professional Roles'. In: S. White (ed.) *Media, Culture and Society in Putin's Russia*, New York: Palgrave Macmillan: 109–132.

Poberezhskaya, M. (2014) 'Climate Change Communication in Russia and New Media', *Digital Icons: Studies in Russian, Eurasian and Central European New Media*, 11: 37–50.

Reporters Without Borders (2012) 'Russia', http://en.rsf.org/russia-russia-12–03–2012,42075. html, date accessed 30/04/2012.

RIA Novosti (2012) 'O RIA Novosti', http://eco.ria.ru/, date accessed 12/03/2012.

Richter, A. (1995) 'The Russian Press after Perestroika', *Canadian Journal of Communication*, 20/1: 7–23.

Sakwa, R. (2011) *The Crisis of Russian Democracy. The Dual State, Factionalism and the Medvedev Succession*, Cambridge: Cambridge University Press.

Schwirtz, M. (2011) '2 Leaders in Russian Media Are Fired after Election Articles', *New York Times* (14 December), www.nytimes.com/2011/12/14/world/europe/russian-journalists-at-kommersant-vlast-axed-after-tough-election-coverage.html, date accessed 1/05/2012.

Shilovskiy, O. (dir.) (2010) *Sgorim. Zamerznem. Vyzhivem*. Documentary film. Moscow: Pervyi kanal.

Shumilina, T. (2010) 'Sotsial'naya Pozitsiya SMI v Epokhu Global'nykh Izmeneniy'. In: S. Konovchenko (ed.) *Sovremennye SMI Rossii: Teoriya i Praktika. Sbornik Nauchnykh Statey*, Moskva: Moskovskiy Gosudarstvennyi Universitet Pechati: 53–67.

Soldner, M. (2008) 'Political Capitalism and Russian Media'. In: S. White (ed.) *Media, Culture and Society in Putin's Russia*, New York: Palgrave Macmillan: 154–177.

Strovskiy, D. (2011) *Otechestvennaya Zhurnalistika Noveyshego Perioda*, Moscow: Uniti-Dana.

Vartanova, E. (2012) 'Russian Media Model in Post-Soviet Dynamics', www.vartanova.ru/articles/en/russian.html, date accessed 12/11/2012.

Vinogradov, D. (2006) 'Russian Internet Remains an Island of Free Speech and Civil Society', *Russian Analytical Digest*, 9: 12–15.

Voltmer, K. (2000) 'Constructing Political Reality in Russia: Izvestiya – between Old and New Journalistic Practices', *European Journal of Communication*, 15/4: 469–500.

Yushchenko, M. (2007) 'Sredstva Massovoy Kommunikatsii kak Mekhanizm Formirovaniya Vlast'yu Obshchestvennogo Soznaniya Grazhdan', www.lib.tsu.ru/mminfo/000063105/305/image/305-067-070.pdf, date accessed 12/01/2012.

Zassoursky, I. (2004) *Media and Power in Post-Soviet Russia*, New York: M. E. Sharpe.

Zasurskiy, Y. (ed.) (2001) *Sistema Sredstv Massovoy Informatsii Rossii*, Moscow: Aspekt.

Zasurskiy, Y. (2004) *Iskushenie Svobodoy. Rossiyskaya Zhurnalistika: 1990–2004*, Moscow: Izdatel'stvo Moskovskogo Universiteta.

Zhukova, N. (2007) 'Sravnitel'nyy Analiz Roli SMI v Sovremennom Politicheskom Protsesse Rossii i SSHA', *Vlast'*, 11: 41–43.

3 Russian climate change policy

Towards 'climate pragmatism'

As argued in the previous chapter, the Russian state has significant influence over media activity. Following this logic in order to understand the media coverage of climate change in Russia, we first need to study the state policy on this issue. Hence this chapter examines Russia's climate change policy in order to conclude whether it has changed since the early 2000s and, if it has, in what ways this can affect alterations in media coverage of climate change.

Russian climate change policy is an ambiguous and complex phenomenon, which can be interpreted in different ways. During the interviews conducted for this research project, the same introductory question which aimed to invite interviewees to discuss Russia's climate policy provoked polar responses: from the straightforward 'it does not exist' to the optimistically sarcastic 'now it does exist, and that is already a positive sign'. Indeed, signing agreements, implementing laws and creating special inter-institutional committees coexist in Russia with very limited practical outcomes and a lack of coordination and persistence, as well as occasionally contradictory policy decisions.

As one of the key controversies, Russian state officials have a history of referring to Russia as an 'environmental leader' or an 'environmental donor'. Russia was labelled an 'environmental leader' in the 1990s because of its drastic, but involuntary, drop in GHG emissions (after the collapse of the Soviet Union) (President of Russia website 2009a). It was referred to as an 'environmental donor' due to its natural geographical resources, in particular the vast areas of boreal forest (which act as a 'carbon sink') (Medvedev 2012, Wilson Rowe 2013). At the same time, national and international environmental communities characterise Russia as an 'anti-leader' of climate change mitigation policy (RSEU 2012) which, along with other countries, including Canada and Poland, was 'honoured' with the 'fossil of the day' anti-award during the Doha Conference (2012) due to its resistance to the negotiation process (Ekoreporter 2012).

Indeed, due to its geographical position, its heavy reliance on the export of fossil fuels and the low energy efficiency of its economy, Russia is now one of the main emitters of GHGs (Parker and Blodgett 2008, Perelet et al. 2007). It has been argued that the environmental situation has worsened by its subordinate position to the state's economic interests (Henry and Douhovnikoff 2008, Porfiriev 1997, Yanitsky 2009), which has resulted in the downgrading of environmental institutions and the concentration of power in a limited circle of the ruling elite (Crotty 2003, Kotov 2002, Oldfield and Shaw 2002, Peterson and Bielke 2001). These factors have contributed to the development of a policy of 'de-environmentalism' (*de-ekologizatsiya*) (Yablokov 2010).

On the other hand, due to the significant drop in its GHG emissions seen in the 1990s and its vast natural reserves, Russia has the capacity to be considered an 'environmental leader' (Klyuev 2002, Tynkkynen 2010) with the ability to influence the world's climate change policy. The Russian government exercised this influence to a great extent during the Kyoto Protocol negotiations (Afionis and Chatzopoulos 2010, Andonova 2008, Buchner and Dall'Olio 2005, Korppoo 2008).

This chapter contributes to the debate outlined above by examining how Russia has hitherto prioritised economic growth over environmental protection. However, it is increasingly in Russia's own economic interest to cut its carbon emissions, which also helps the country to promote its global integration. Based on content analysis of 72 presidential speeches made by Medvedev between 2008 and 2012, it is argued that the governing elite, especially Medvedev, has started to recognise the economic benefits of Russia's proactive climate policy. These economic considerations are driving Russian climate policy in two ways: the direct benefits from mitigation (for instance, by improving energy efficiency) and the indirect benefits from integration with the global community (for example, bringing 'green' investments into the country or portraying Russia as a modern trade partner that cares about its 'green' image). Hence, with regard to Russia's climate policy we can witness the evolution of *climate pragmatism*, where the state can see the benefits from remaining 'faithful' to its economic development plans and at the same time becoming a 'real' environmental leader that actually cuts its emissions rather than relying on a fortuitous drop in emissions, as in the past.

This chapter first looks at the evolution of Russia's climate change policy through the perspective of two competing views: firstly, Russia as a 'de-environmentalist' and secondly, Russia as an 'environmental leader'. Then it explores the new course in climate policy in Russia which coincided with Medvedev's presidency – hence, through the analysis of his official speeches, this chapter looks at how the new emphasis on economic modernisation has become beneficial for climate policy.

Between 'de-environmentalism' and 'environmental leadership'

The concept of 'de-environmentalism' or 'de-ecologisation' (*de-ekologizatsiya*) has been popularised by the former special adviser to President Boris Yeltsin on environmental and public health affairs and a current chairman of the Green Party faction of the political party 'Yabloko', Alexey Yablokov. He states that 'Russia's environmental problems are the result of the state policy of "de-environmentalism", where dealing with environmental problems is postponed until the country reaches a certain level of wealth, and until then it serves as a "reservoir of natural resources"' (Yablokov 2010, p. 3). Yablokov argues that this approach began to develop under Yeltsin and advanced during the time of Putin (first two terms) and Medvedev. He identifies several stages which constitute the process of 'de-environmentalism', among them 'the dissolution of the Environmental Protection Agency [. . .], a weakening of environmental protection legislation [. . .], the obstruction of environmental NGOs [. . .], a reduction of funding for environmental programmes' and so on (ibid, p. 4). Even though Yablokov does not specify the particular area of environmental problems which could be characterised by the phenomenon of 'de-ecologisation', he explores the country's general environmental degradation (starting with air pollution and finishing with public health problems associated with it). It can be argued that this concept accurately describes Russian climate policy.

Another leading Russian scientist, Nikolay Klyuev (2002), states that the academic and public debate emerged at the end of the 1980s (to talk about Russia's ecological situation in a highly pessimistic way), lacks any foundation and damages the country's investment and recreational appeal. Furthermore, he argues that Russian territory is the main natural purification system of the planet – it compensates global pollution and overall acts as an environmental donor (ibid, p. 19). Vladimir Zakharov (2011, p. 6) shares a similar opinion – '[because of the recent] economic growth, rich natural resources and the search for ways of optimal development [. . .] [t]his will make it possible to rank Russia not only as an energy power but also as an environmental donor.' In Klyuev's (2007) later work on the comparative analysis of states' 'eco-industrial pyramids' (the correlation between a country's industrial capacity, resources consumption and waste production), he concludes that in order to become an environmental leader Russia still needs to improve its industrial production process. The argument of Russia's environmental leadership was promoted by Russian officials and covered by the media during the Kyoto Conference in 1997 to the extent that it was claimed that 'the Russian delegation performed a diplomatic miracle' (*Izvestiya* 1997) and led the negotiations (see more in Chapter 4). As previously mentioned, Nina Tynkkynen (2010) suggests that the media exploitation of the 'Great Environmental Power' concept in the coverage of Russia's climate change policy diverts attention from Russia's resistance to the Kyoto Protocol's ratification and the carbon intensity of its economy. Instead it highlights Russia's drastic drop in GHG emissions and its natural capabilities (due to its forests) to solve climate change problems – 'a source of environmental solutions rather than a source of environmental problems' (ibid, p. 182).

The outlined controversies in Russia's climate policy to some extent are embedded in its geographical specifications, where, as discussed below, the wealth of natural resources coexists with extreme vulnerability to the effects of climate change.

Russia's geography and climate change consequences

Russia's rich natural resource reserves and vast territory define the state's economic orientation and its influence on the global environmental situation. In fact, even after the breakup of the USSR, Russia remains the largest state in the world, containing such different climate zones as arctic, sub-arctic, temperate and sub-tropical (Perelet et al. 2007, Shaw 2009). Russia holds the record for the maximum temperature in the world: 116.6°C. Natural zones vary drastically from polar desert, tundra, taiga, mountains and mixed forest to steppe and semi-desert. Another key geographical characteristic is Russia's leading position in reserves of natural resources such as natural gas, oil, coal, iron ore, bauxite, nickel, tin and so on (Orlenok et al 1998). Significantly, most of these resources are situated predominantly in permafrost (which covers 65 percent of Russia's territory) and in severe climatic conditions, leading to the high cost of their extraction and transportation.

There is the possibility that the large size and geographical nature of Russia's territory, the diversity of climate zones and the location of natural resources, as well as the country's very low population density, might lead to positive consequences from climate change for Russia. For instance, the majority of Russia's territory is situated in the area of maximum warming, and so the softening of climate conditions could extend the zone of 'comfortable living' to the northern border, reduce energy expenses during the heating season, facilitate access to natural resources, prolong harvesting seasons, decrease cold-related illnesses and deaths, improve transportation through the Arctic seas and facilitate the development of the Arctic shelf (Kattsov et al. 2007, Perelet et al. 2007; see also in Fay et al. 2010). This tends to promote a false optimism among many Russians about climate change. However, lately more and more people in government and science have come to the realisation of climate change's damaging character for Russia's ecosystems, economy, security, infrastructure and so on. In this case, the climate zone diversity of the country is considered a weakness rather than a strength. Renat Perelet, Sergey Pegov and Mikhail Yulkin (2007) in a Human Development report describe in detail what the vulnerabilities of each zone could be. For example, the most fertile regions will suffer from droughts. Russia's famous forest zones such as taiga and tundra will shrink and be exposed to outbursts of forest diseases. The steppes will also experience more droughts, loss of harvest and replacement by other ecosystems. Deserts will suffer from increases in strong winds and storms.

It has been concluded that climate change is taking place quicker in Russia than in the rest of the world (Charap 2010). From 1907 to 2006 the global average temperature rose by 0.74°C, whilst in Russia it increased by 1.29°C (Bogdan et al. 2009). The latest Russian *Assessment Report of Climate Change and Its Consequences* (Roshydromet 2014) has reaffirmed this tendency, stating that since the mid-1970s average temperature in Russia has been increasing by 0.43°C, which is two times faster than the average speed of the global temperature rise. The impacts of this rapid temperature change can be readily observed (Bogdan et al. 2009). For example, 2010 was an extreme year which exceeded the 'normal' temperature (the norm being 1961–1990) by 0.65°C. Even though it can be characterised as only slightly anomalously warm, it consisted of an extremely cold winter, an extremely hot summer and an extremely warm autumn (Kattsov et al. 2011).

These extremes led to severe consequences for Russia's economy, for nature and for people's health. Moreover, some areas of Russia are more vulnerable to temperature increases than others – as climatologist Nataliya Kharlamova stated, without the moderating factor of the sea, climate change is particularly apparent in the Altay region, Tyvy and next to the borders with China and Mongolia (interview 13, February 2012).

Another fact stressed by climatologists is that climate change in general happens quicker in the polar territories, which means that Russia, with its large proportion of permafrost, might see widespread melting, which could lead to severe economic and social damage (most of the oil and gas industries are located in this type of territory, for example[1]; for more, see Gotz 2007). By the start of 2014, it had been reported that 60 percent of 'objects of infrastructure' in Igarka, Dikson and Khatanga were deformed, 100 percent of 'objects' in villages in the Taymyr Autonomous Okrug, 22 percent in Tiksi, 55 percent in Dudinka, about 300 structures in the area of Norilisk and so on (Roshydromet 2014). Temperature increases will also make it impossible to transport timber by 'winter roads', which will lead to more expenditures on building new routes (Roshydromet 2008).

This introduction to the geography of Russia highlights the factors which initially put the country into an ambiguous position where resource wealth exists parallel to severe weather conditions. This, along with industrial production and other economic activities, has resulted in Russia being the third biggest CO_2 emitter in the world after the United States and China (if the EU is considered, then Russia is fourth) (CRS 2008) – so the way Russia deals with this complex situation and governs its assets naturally makes it one of the leading countries influencing global climate policy.

The next specification that needs to be considered in the analysis of Russia's climate change policy, is the country's administrative structure, which explains not only the heterogeneous attitude towards the problem amongst various subjects of the state, but also the supremacy of Russia's executive branch.

Russia's administrative structure and climate policy

Russia is a federation consisting of 83 federal subjects with different status and degrees of autonomy (Constitution of the RF 1993). The importance of the acknowledgement of the federal structure of Russia goes along with the consideration of any particular region's distribution of powers to enact and implement its budget and laws. Depending on the status of the particular region (whether it is a republic or an oblast), it will have a certain degree of autonomy from the federal budget and laws. Moreover, authorities at the regional level might be more aware of the ecological problems in the area. Some regions are heavily populated by fossil fuel industries, some areas are more vulnerable to the impact of climate change, and so on (Firsova and Taplin 2009). This regions' diversity to a degree influences their support for climate change policy, which was demonstrated by the WWF survey conducted prior to Russia signing the Kyoto Protocol (Buchner and Dall'Olio 2005). The central and European parts of the country (especially the northwest) were mostly pro-Kyoto, whilst more remote areas of Siberia demonstrated a lack of support.[2]

The federal structure of the country has also influenced its legislative branch. The Russian Constitution (1993) established divisions between areas of federal and regional jurisdiction and confirmed the priority of federal law in areas where they overlap. Environmental protection legislation falls under joint jurisdiction (Buchner and Dall'Olio 2005). In this case the blurred boundaries between federal and local responsibilities could really damage the development of environmental protection actions, as occurred after the ratification of the Kyoto Protocol. As a regional environmental activist admits: 'frankly speaking even though we now have a presidential adviser on climate change, it did not change much here in our region. I am not even sure who is in charge of this problem. I suspect that it is spread throughout different departments and it has become (for them) just another line in the report' (interview 26, Barnaul, 9 August 2011).

Laura Henry and Lisa Sundstrom (2007, p. 62) have pointed out that after the ratification of the Kyoto Protocol, 'the fundamental question of property rights' was not defined. It was not clear who would be allowed to trade carbon emissions and get the profit if Russian businesses stayed below the permitted level: the federal or regional governments, or maybe business. One of the interviewees mentioned that local authorities would be very interested in the benefits the Kyoto Protocol might bring for the region but that due to the absence of a legislative base, nothing could be done. There was also a fear that without working laws on the subject matter, it might lead to abuses of power and corruption. So, in the case of the Kyoto Protocol, the necessary legislation at the domestic level was not signed until 2007, then was changed in 2009, which delayed approval of the first JI project[3] in the country until July 2010 (Henry 2010).

Even if legislative problems are clearly solved through different laws at the federal and regional levels, another peculiarity of the state's legislative system is that Russia has very strong presidential powers (and when Putin exchanged this post for the prime minister position [2008–2012], it had a very powerful prime minister too). It means that though the Federation Council is supposed to represent all subjects of the federation on legislative matters, the president has the power to 'unilaterally overturn regional acts and laws in his role as a protector of the constitution' (Buchner and Dall'Olio 2005, p. 363).

Interestingly enough, many NGO representatives referred to Putin's 'special attitude' towards climate change problems. One of them mentioned that Putin did not even like the word 'Kyoto' (interview 19, Moscow, July 2011), whilst another environmentalist claims that while Putin is in power, climate policy will remain in its infancy: 'whatever Medvedev says, it is just plans which are not getting fulfilled, while for Putin, Gazprom and Rosneft are his interests, which he will never abandon' (interview 6, Moscow, July 2011).

Unfortunately, even when other actors express direct interest in climate change affairs, in most cases it comes down to the decision of the country's leader or leaders. Moreover, even the president's adviser on climate change, Alexander Bedritsky, just before the UNFCCC in Qatar, stated that the decision of Russia's position on the continuation of the Kyoto Protocol was made at the 'highest political level' and could be changed only by the president himself (RIA Novosti 2012b). Dobrovidova (2012b) goes so far as to argue that Russian experts on climate change are 'in essence taking part in a ceremony' and all of the debates around Russia's climate policy do not really affect the final decisions, or as Dobrovidova (ibid) states, 'in climate policy terms – discuss all you want, as the real decision makers are as far from the debate as Europe is from New Guinea'. This pessimistic picture of the fate of climate change concentrated in one set of hands is worsened by the cumbersome structural changes of the environmental institutions and the state's dominance over the energy sector.

Institutional change: Ministerial paradox

When the new state of the Russian Federation was created in 1991, it inherited a poor environmental record, which had led to the degradation of the environment in Russia (Feldman and Blokov 2009, 2012). Even though in the early 1990s the idea of environmental protection grew and among other concepts became quite fashionable, it was also employed as cover in order to recreate the country's image and position internationally as a liberal country which followed modern world trends. The euphoria did not last long, as is shown by the institutional change in environmental protection that the state has experienced over the short period of time since the 1990s (Henry and Douhovnikoff 2008).

Some argue that this transformation and degradation of the relevant environmental institutions became one of the stages in the development of the 'de-ecologisation' policy (Oldfield 2001, Yablokov 2010, Yanitsky 2011). However, looking at the changes happening within the Soviet and Russian environmental institutions, the concept of 'de-ecologisation' raises a number of questions. For instance, the term suggests that Russian policy was at one time 'ecological', but this idea of the superiority of economic development over the environment and the use of nature as a means to achieve the state's goals is very much a Soviet concept[4] (Feldman and Blokov 2009). As Yanitsky[5] (2009, p. 754) states, 'until the 1970s, the dominant worldview rooted in Soviet culture was strictly utilitarian. Nature was seen as an unlimited resource pool that had to be (re)constructed in accordance with the goals of the construction of a socialist society' (see more in Oldfield 2005). In his extensive research of the shift in environmental debate in Russia, Yanitsky (2009) states that this utilitarian approach to nature in the mid-1980s to early 1990s was replaced by the 'greener' idea of Russia as being a 'limited space [which] must be kept clean and safe'. From the mid-1990s and early 2000s the concept of Russia as an area of 'unlimited resource' reappeared in the public space. The findings of Yanitsky's research discussed above and the following discussion of institutional change lead us to the conclusion that the so-called policy of 'environmentalism' could only be traced back to the time of Gorbachev's policy of *glasnost* and *perestroika*.

Historically, in the USSR, environmental issues were the responsibility of different ministries, so there was not one legal authority responsible for environmental

protection. It was only in 1988 that the State Committee on Environmental Protection (Goskompriroda) was founded. Many saw this as a governmental response
to the recent environmental catastrophe at the Chernobyl nuclear power plant in
1986 (Henry and Douhovnikoff 2008). After the collapse of the Soviet Union, for
a short while environmental protection advanced in the hierarchy of the political
agenda as well as the Goskompriroda, which in 1991 became the Ministry of
the Environment. Already by 1996 the importance of environmental protection
was downgraded and the Ministry became the State Committee on Environmental
Protection (under a slightly different name – Goskomekologiya). In 2000 even
this committee was dissolved, whilst some of its functions were transferred to the
Ministry of Natural Resources (Henry and Douhovnikoff 2008, Oldfield 2001,
Peterson and Bielke 2001). Finally, in 2008 the Ministry of Natural Resources
became the Ministry of Natural Resources and Environmental Protection, and it
has remained the main authority in the country which at the state level deals with
environmental problems.

The dissolution of the Goskomekologiya was perceived as a partially positive
decision, as it lessened the amount of bureaucratic obstacles in the way of environmental management (Kotov 2002) as well as giving to one organisation the
opportunity to have expertise and responsibility on environment-related issues[6]
(Firsova and Taplin 2009). On the other hand, the fact that its functions were
transferred to the Ministry of Natural Resources weakened the state's domestic
policy towards environmental protection (Firsova and Taplin 2009, Oldfield and
Shaw 2002) and stimulated to an even greater degree the growth of an economy
heavily based on the exploitation of natural resources (Henry and Douhovnikoff
2008). When the final institutional transformation took place, the former chair
of the Goskomekologiya, Professor Viktor Danilov-Danilyan (who along with
Yablokov claims that Russia is pursuing a policy of 'de-ecologisation') declared:

> [it] is a signal for thieves. The law says, 'Hey guys, there is no one watching
> over nature, so come and take what you want! [. . .] Authorising the Natural
> Resources Ministry to deal with environmental problems is like asking an
> alcoholic what the price of vodka should be'.
>
> (Peterson and Bielke 2001, p. 69)

The fate of domestic climate change policy was also influenced by the confusion between institutions of environmental protection and the ones which would
somehow be responsible or concerned with it. As the head of the department of
sustainable development and partnership at the Sustainable Energy Development
Centre, Vladimir Berdin, noted, 'any changes in such institutions lead to the temporary stagnation of all processes; it takes time for people to distribute responsibilities and get back to routine work' (interview 4, Chemal, 13 August 2011).
Even at the time when the Ministry of the Environment and later the Goskomekologiya still existed and had some powers, climate change policy had involved
various ministries and interest groups. In order to manage relationships between
different institutions on the issue, the Interagency Commission of the Russian

Federation on Climate Change was established in 1994 (Climate Change Action Plan Report 1999). The commission united representatives from 21 ministries and interest groups such as the Ministry of Foreign Affairs, the Federal Service for Hydrometeorology and Environmental Monitoring (Roshydromet[7]), Gazprom and so on (ibid, p. 15). The commission was supposed to be the supreme authority in deciding climate change issues. At the same time, it did not have any legal authority and could not regulate concrete projects or deal with investments. Its powers were weakened when the Goskomekologiya was abolished. Buchner and Dall'Olio (2005) argue that first, Goskomekologiya had a much bigger interest in climate change policies than its successor, the Ministry of Natural Resources, and, second, Goskomekologiya simply had more expertise and resources on the topic.

Eventually, the Interagency Commission proved to be inefficient, and even though officially it was supposed to be the main approval body for documents on climate change regulations, its functions became part of the Roshydromet. The commission did not meet at all for several years and then was dissolved. Vladimir Kotov (2002) has commented on the inefficiency of the commission that '[it] was a typical institution of the transitional period: old form but without old possibilities, a remainder of the old system not adapted to the new institutions' (ibid, p. 16). The idea of inter-ministry and inter-agency collaboration is still very popular and discussed by the scientific community as well as by representatives of the government and NGOs. For example, Kattsov et al. (2007) give examples of how exactly different ministries and state agencies will be interested in dealing with issues connected with climate change (see Table 3.1).

To sum up, the dissolution of the State Committee on Environmental Protection and the transfer of some of its functions to the Ministry of Natural Resources led to 'an institutional paradox' in which the institution responsible for the exploitation of natural resources also had to be in charge of its protection. It leads back to the theoretical paradox discussed above, where the state's resource wealth and capabilities of becoming an 'environmental donor' are undermined by the downgrading of the environmental institutions, which once again contributes to the development of a policy of 'de-ecologisation'. At the same time, the multifaceted nature of climate change problems and the interests of different state organisations present an opportunity (rather than a burden) for inter-institutional collaboration, not only at the state but at the international level as well. Realising and using the benefits of climate change collaboration will not be possible without a consideration of Russian economic interests and in particular the peculiarity of its energy sector, which also works in both directions: heavy reliance on natural resources stimulates the degradation of the environment, but the vast amount of natural resources gives Russia power in international discussions of climate change regulation.

The role of the energy sector

At the beginning of this chapter, the importance of economic interests was listed among the major factors which affect Russian climate change policy. Over the

Table 3.1 Examples of possible climate change-related problems which might concern different Russian federal institutions

Ministry	Examples of Interests Related to Climate Change
Ministry of the Interior	Migration processes
Ministry for Civil Defence, Emergencies and Elimination of the Consequences of Natural Disasters	Increase in natural disaster frequency
Ministry of Foreign Affairs	International agreements on environment and climate change
Ministry of Defence	Defence of the borders in relation to changes of the geopolitical situation and, in particular, defence of Russia's sovereignty in the Arctic region
Ministry of Health and Social Development	Climate change-related threat to the population's health
Ministry of Education and Science	Preparation of qualified scientists to serve Russia's interests
Ministry of Natural Resources	Climate change's negative impact on natural resources; easier access to the resources in the Arctic and as a consequence its environmental pollution
Ministry of Energy	Problems of energy saving, alternative energy sources, sustainable technology. Monitoring of GHG emissions
Ministry of Regional Development	Climate change impact on regional economy and local infrastructures such as the melting of permafrost
Ministry of Agriculture	Change in harvest, land used for agriculture, fishing and new types of parasites
Ministry of Transport	Development of the new Arctic route. Impact of melting permafrost on motorways and railways
Ministry of Information Technology and Communication	Participation in IT support for climate research and monitoring of the climate
Ministry of Finance	Financing the priority scientific research on the climate
Ministry of Economic Development and Trade	Economic justification of political and economic decisions on climate change issues

Source: Adapted from Kattsov et al. 2007.

decades, 'the legacy of the conservative command-and-control processes' and 'the transitional state of development in Russia' contributed to the prioritisation of economic growth over environmental protection, which relegates environmental protection to the background or postpones dealing with it until 'better times' (Porfiriev 1997, p. 148).

The wealth of natural resources in the country (especially gas, coal and oil) carried on playing a significant role in the transition of Russia's economy after the collapse of the Soviet Union. Russia has the largest reserves of natural gas in the world, and it heavily depends on these for domestic consumption and as an export commodity.[8] Sergey Aleksashenko (2012, p. 43) argues that the Russian economy is 'de facto monocultural', since 'more than 85 percent of Russian exports are either raw materials or primary commodities' and 'during 2010–2011 the share of hydrocarbons in Russian exports fluctuated between 63.5 and 65 percent'. Russia's resource-oriented economy is reinforced by its carbon intensity (the amount of GHG emissions per unit of GDP), which 'exceeds the leading European countries by 3.8 times, the average for transition economies by 2.6, the USA by 2.4 and Canada by 2 times'[9] (Perelet et al. 2007, p. 10), making Russia one of the most polluting economies.

The significance of the energy sector in domestic policy on climate change should also be considered from the perspective of the close connections between the energy sector and the state. Nowadays, major gas, oil and electricity companies are either partly owned by the government directly or by entities which are close to the Kremlin.[10] It was not always like this, and for a short period of time private ownership prevailed (Buchner and Dall'Olio 2005). For instance, during the early 1990s the oil and gas industries were privatised, and this era in the country's history is famous for the rise of the 'new class' (the 'oligarchs'). After acquiring fortunes through monopolising the state's essential infrastructures, very soon some of these oligarchs crossed the line by not focusing as much on reinvesting profit into their businesses. Instead they avoided paying taxes and began 'moving cash offshore' (ibid). The era of Putin's government was characterised by the policy of centralising the energy sector and increasing the state's influence over it.

At the time of writing, the major actor in the energy sector is the state. It owns all shares in the second largest oil company, Rosneft. Another oil company, LUKoil, has a mostly private ownership structure where only 14 percent of shares belong to the government, but it retains close connections with the government. Buchner and Dall'Olio (2005, p. 365) point out that 'individual companies have very different relations with the state that do not always reflect the state's share in the enterprise.' Overall, the state is in charge of 30 percent of oil production in the country; however, if all informal connections are considered, that figure might be higher. Gas production in Russia is dominated by Gazprom, of which approximately 51 percent of shares belong to the government. It is not only Russia's largest gas company, which controls 90 percent of domestic gas production (ibid), but it is also one of the world's dominant players in the energy sector and has become one of the most powerful tools in Russia's foreign policy. As Sharples (2013, p. 686) has highlighted, 'for Russia, natural gas is not only an economic resource. Energy security is perceived to be a political concept'.

The role of the energy companies in Russia's climate policy is quite ambiguous. For instance, Henry (2010, p. 767) highlights the aspect of businesses' reluctant attitude towards environmental issues by quoting some Russian environmental activists: 'most commercial firms do not want to be associated with "democracy, human rights or the environment" and the wealthiest firms are "too dependent on exploiting natural resources" to give funding for environmental causes'.

According to a climate change activist who has approached (and has been approached by) a number of energy companies during his NGO's energy efficiency campaign (interview 31, 27 July 2011), business firms clearly understand that climate change is not a 'PR campaign' but a serious science. At the same time, they understand that they cannot refuse to follow the government's orders, so they are trying to sabotage the fundamentals of it, by popularising the idea that there is no anthropogenic cause to climate change. In Russia there are no threats to business coming from the climate change policy (no laws restricting them or judgemental public opinion) and because of it, they are not very active. However, they all have complete information of the problem, and some companies even have calculated their emissions[11] but 'sit quietly' so long as it does not directly concern them. Another climate change campaigner (interview 19, 27 July 2011) shares his experience working with one of Gazprom's companies: '[t]hey openly talked to us about the problem and said that they already include climate change costs into their projects, because their infrastructure is based on permafrost, and changes have already forced them to reinforce their buildings and structures'. His colleague (also coming from his work experience) in support of this argument adds that Gazprom managers understand the danger of climate change for their business, and their scientists tell them that climate change might make their project unprofitable, but Gazprom people do not want to discuss this topic in public in order not to diminish the value of their shares on the market (interview 6, 27 July 2011).

Indeed, Gazprom's deputy chairman of the board of directors, Alexey Miller, keeps referring to climate change as having been created by the media as a 'PR campaign' (Mason 2011). At the same time, this large gas company, according to Buchner and Dall'Olio (2005), was supporting the ratification process of the Kyoto Protocol and was 'keen to maintain a green image'. For instance, on Gazprom's official website, one of the pages is solely devoted to the Kyoto Protocol, explaining how Gazprom complies with its requirements and overall how 'Gazprom Group makes consistent efforts to reduce GHG emissions' (Gazprom 2014). Once again we see a paradoxical situation: as much as Russia's economy is heavily based on exploitation of fossil fuels, the companies involved in these industries could treat climate change regulations not as a threat to them, but as a profitable venture and a way to attract more investments and modernise their production processes[12] (Mandrillon 2008, RIA Novosti 2012a).

The vast reserves of natural resources that Russia possesses can be seen as a double-edged sword. Overdependence on them in the economy contributes to Russia's leading position among GHG emitters. On the other hand, the contraction of its economy in the 1990s gave Russia significant status during international negotiation processes. As Oldfield and Shaw (2002, p. 392) state: 'uncertainties about Russia's future must translate into uncertainties about the future well-being of the global environment in general'. The way the factors outlined above influence Russia's climate policy is evidenced by the case study of Russia's involvement in the Kyoto Protocol ratification process. For six years the official Russian position on Kyoto ratification swung back and forth, demonstrating the inconsistencies in the government's agenda.

The Kyoto Protocol negotiations, or Russia's 'environmental blackmail'

The Kyoto Protocol (1998) was the first document which forced signatory indus-trialised countries to 'commit' to certain GHG emissions obligations. It also acknowledged the developed countries' historical responsibility for the current accumulated amount of GHG in the atmosphere, and applied more rigorous restrictions to them. Even though the protocol was adopted on 11 December 1997, it only came into force on 16 February 2005. Strangely enough and quite unexpectedly at that time, Russia became one of the key reasons for the delay in the agreement's implementation (Tipton 2008).

Due to the conditions under which the protocol could enter into force, it had to be ratified by at least 55 countries, and the participants had to be responsible for at least 55 percent of global GHG emissions (Kyoto Protocol 1998). When the United States refused to sign the agreement, stating that the protocol would damage its economic interests, the fate of the document ended up in the hands of the Russian government, since with its contribution the percentage of GHG emissions the protocol covered would attain the required level. One might argue that during the time of the Kyoto negotiations, the actual 'ecological power' of Russia could be seen particularly clearly, as well as how it could be converted into political power utilised by the Russian government in its own national interests (Buchner and Dall'Olio 2005).

During the difficult years of the early 1990s, the new state of the Russian Federation tried to reestablish its role in the new world order. It went through the period of committing itself to numerous international agreements and trying to join various international institutions. In the sphere of environmental cooperation Russia signed '30 bilateral environmental agreements and [joined] more than 25 regional environmental regimes' (Henry and Douhovnikoff 2008, p. 451). In 1992 Russia was among the first countries to sign the United Nations Framework Convention on Climate Change and ratified it in 1994. Russia was classified as an 'economy in transition' and ratification did not imply any obligations. Very shortly after the convention came into force, Russia changed its position by raising concerns over climate change mitigation's impact on its energy policy. Furthermore, it took the side of the Organization of Petroleum Exporting Countries (OPEC), which questioned the limits set by the UN convention (Andonova 2008).

By the time the conference in Kyoto, Japan, took place in 1997, Russia had left the OPEC camp and was already in the camp with the United States, Canada, New Zealand and Japan, who lobbied for lower emission restrictions for industrialised countries. When in 2001 the United States announced its withdrawal from the Kyoto negotiations, Russia faced two outcomes. On the one hand, it lost the largest potential buyer of spare emission quotas (Oldfield et al. 2003). On the other, Russia suddenly gained greater negotiating power due to its possible impact on the protocol's fate. Afionis and Chatzopoulos (2010) argue that the EU realised straight away that in order to make the Kyoto Protocol a reality, it had to comply with the demands of Russia, Japan, Canada and Australia (after the United States and China, the world's largest GHG emitters). Eventually in 2001 at the conference in Bonn, the EU offered such compromises that it became impossible to say no without damaging a country's international reputation. One of the conditions was to consider 'sinks' towards the estimation of the state's GHG emissions. For example, countries with massive forest zones (such as Russia), would be considered less polluting, since a certain amount of GHG emissions would be sunk by their natural carbon-absorbing reservoirs.

The wave of enthusiasm in Russia that followed the Kyoto Protocol's ratification soon disappeared, when President Putin radically changed his position once again and asked for agreement on certain sums to be invested into emissions trading or JI mechanisms. In addition, a few years later at the COP-9 (Milan 2003), Russia managed to announce during a very short period of time, first, its firm intention not to ratify the protocol, and second, that ratification was still very much under consideration and the country was moving towards it (Buchner and Dall'Olio 2005). It took another year before Russia finally signed the protocol, on 18 November 2004.

As demonstrated above, before Russia signed the protocol, it achieved certain political and economic guaranties. Andonova (2008, p. 489) argues that Russia managed to maintain such 'considerable bargaining power' because its participation was vital for the protocol to come into power, whilst Russia did not have any national interests in the climate change negotiations as the country's economic development was the top priority. So, Russia stated many times that it would not commit to any other targets which were under the 1990 level. It must be explained that compared with the post-*perestroika* years, in 1990 Russia was still in its peak period of industrial capacity, which was accompanied by a high amount of GHG emissions. After the collapse of the USSR, the country went through major economic decay, which consequently decreased its GHG emissions by approximately 40 percent[13] (Andonova 2008, Oldfield 2005).

Another explanation of why it took so many years for Russia to sign the Kyoto Protocol arguably is that the protocol was opposed by a group of very influential scientists and economists who were not so sure of the human contribution to climate change and its negative consequences for Russia, whilst they were convinced that the protocol would restrict the economic growth of Russia. Two in particular, due to their positions and authority, had been the most vigorous opponents of Kyoto's ratification: Yuri Izrael, a former scientific adviser to President Putin and a director of the Russian Academy of Sciences' Global Climate and Ecology Institute, and Andrei Illarionov, at that time the president's chief economic adviser. Izrael was (until 2008) also a vice-chairman of the IPCC. However, his opinion of Kyoto's useless and damaging character for the Russian economy persisted even after Russia's ratification – he asked the president to annul his signature. Illarionov, refusing to acknowledge the anthropogenic character of climate change, referred to ' "Kyotoism" as a new "totalitarianism" and said that its implementation would be an "Auschwitz for civilisation" ' (Mandrillon 2008, p. 135, citing Rosbalt 2004). Two years before the ratification took place, Illarionov announced an economic model in which he projected the doubling of

Russia's GDP, which would lead to a situation in which Russia would exceed the GHG emissions limit prescribed by the protocol and would be forced to buy quotas (Buchner and Dall'Olio 2005). After the protocol was ratified, Illarionov,[14] together with Natalia Pivovarova (director of the Economic Analysis Institute), published an article on the economic consequences of the ratification (Illarionov and Pivovarova 2004). They stated that the risks and danger of the Kyoto ratification for Russia had become reality and that the government had to deal with them. The authors supported their statement by looking at the correlation between such positions as the amount of financial resources which Russian businesses might receive from emission trading, the amount of money that Russian companies would need to spend to meet the quotas (and fines for exceeding them), the amount of resources needed to respond to the protocol's requirements, and the slowing of economic growth, which would be unavoidable in order to meet the protocol's requirements.

Sergey Kuraev, from the Russian Regional Environmental Centre, does not concentrate on the role of these two prominent scholars, but rather argues that the general decline in Russian science and in particular in the research area of climate change was crucial to the state's policy. Kuraev (2011) says that when in 1992 Russia signed the United Nations Framework Convention on Climate Change, it took only two years to ratify it at the national level, which was possible due to Russia's scientific community, who provided all the necessary scientific information on climate change's impact on people's health and the country's economy, ecosystem and biodiversity. After the Kyoto Protocol was adopted and was waiting for Russia's signature, the academic community took firm anti-Kyoto positions and it became, according to Kuraev, almost fashionable among scientists to say something against its ratification. Among the reasons he names are personal preferences, academic disputes, organisations' rivalries and the decline in the number of new enthusiastic academics in the field who could have changed the trend.

Peter Haas (1989), in his study on the role of epistemic communities in the implementation of Mediterranean pollution control, states that 'epistemic communities may introduce new policy alternatives to their governments, and depending on the extent to which these communities are successful in obtaining and retaining bureaucratic power domestically, they can often lead their governments to pursue them' (ibid, p. 402). It could be argued that due to the high positions of these two Kyoto opponents in Russia (Illarionov and Izrael), their opinion was often perceived as the Kremlin's official position (Henry and Sundstrom 2007). On the other side of the domestic debate of the protocol's ratification were environmental NGOs, other representatives of scientific communities (some argue it was a majority[15]) and in fact some of the biggest businesses in the country. Even though some companies such as Yukos and Norilsk Nickel were against it (fearing that it would restrict the development of their industries in the Arctic), companies such as United Energy Systems, Russian Aluminium, Gazprom, the Russian Union of Industrialists and Entrepreneurs and others admitted the advantages of ratification and supported the protocol (due to the prospects for foreign investments in these industries through the JI mechanism) (Henry and Sundstrom 2007). Russia's ministries opposed each other on different sides of the debate as well. The Ministry of Energy saw it as a way to bring money into the modernisation of the energy sector (Buchner and Dall'Olio 2005). The Ministry of Natural Resources feared the protocol would restrict the use of natural resources. Roshydromet, even though it was affiliated with the Ministry of Natural Resources, was mostly pro-Kyoto, and the Ministry of Economic Development and Trade changed its opinion from negative to neutral when it realised that the procedures required by the protocol could be managed by the Ministry of Energy. However, as Henry and Sundstrom (2007) discovered after the president signed the Kyoto Protocol, the ministries united in support of his decision. They concluded that 'bureaucratic battles among ministries, parliamentary debates, and regional interests are less important than the overwhelming power of the executive branch' (ibid, p. 56).

Overall, when the protocol was signed and ratified, Russia managed to achieve favourable conditions out of the agreement, under which it was not obliged to do anything, as it was very unlikely that it would reach the level of emissions at the 1990 benchmark. Furthermore, it got the opportunity to sell its spare quotas and bring investments into the country through the JI mechanism. And last but not least, Russia's final decision to sign the protocol coincided with the EU's support for Russia's application for World Trade Organization (WTO) membership[16] (Andonova 2008, Henry and Douhovnikoff 2008).

The process of Russia ratifying the Kyoto Protocol demonstrates how climate change policy was moved to the background of the state's other political and economic interests, and the 'climate card' was used when necessary. As discussed earlier, the rebirth of the concept of Russia as a 'world power' was mentioned by several scholars during the discussion of Russia's behaviour at the Kyoto negotiations (Afionis and Chatzopoulos 2010, Henry and Sundstrom 2007). However, since Russia's ambiguous strategy during the Kyoto negotiations managed 'to reduce the credibility of the whole country in the international arena' (Korppoo 2008, p. 7), the Copenhagen Conference could be seen as a second chance for the country to use its climate change policy to rebuild its image as a modern liberal state.

A new chapter in Russian climate policy?

If the Kyoto negotiations brought Russia into the spotlight and made it one of the key players in climate change politics, Anna Korppoo (2008, p. 7) argues that during the post-Kyoto negotiation period, Russia became a 'reluctant party'. After signing the protocol, Russia partially lost the attention it had previously enjoyed due to new actors coming into play: the United States with a new administration, along with China, India and other newly industrialising countries. On the other hand, the Kyoto Protocol's rather tolerant requirements of Russia were not likely to be maintained under a new agreement where, in order to comply with new GHG emissions reduction goals, Russia might have to actually reduce its emissions intentionally through specific policy mechanisms. Korppoo argues that

Russia's government would continue to insist on 'differentiated responsibilities' (ibid, p. 7) and lobbying to be categorised as an emerging country so that its emissions would not be restricted until it reached a certain level of development.

At the Copenhagen Conference,[17] 'global but differentiated responsibilities' became one of the main messages in the Russian president's speech. However, once again, Medvedev talked about Russia as the *leader* among the countries reducing their emissions. Even though in large part it was just a facade (Russia's carbon emissions reductions were still massive, but not due to any specific measures, but rather to economic problems in Russia), it demonstrated the way the government wanted Russia to be presented. Medvedev also announced that regardless of the outcome of the conference, Russia would commit to the 25 percent emissions reduction rate by 2020 (baseline year is 1990) (President of Russia website 2009a).

At the national level the day before Medvedev's speech at the conference, the Russian Climate Doctrine (2009) was adopted. The doctrine acknowledged the importance of anthropogenic influence on climate change and analysed the risks it might bring for Russia. It states that the consequences of climate change can be seen at global, regional and national levels. Global climate change creates a situation which demands a planned governmental strategy for managing climate change problems (especially considering Russia's geographical characteristics, climatic diversity, economic structure, demographic problems and geopolitical interests). The Climate Doctrine states major principles in Russia's climate policy, such as the orientation of Russia's national interests in relation to climate change, acknowledgement of the necessity of international partnership in scientific research, precaution in planning and implementing measures on protecting humans, the economy and the state from the undesirable consequences of climate change and a clear and open information policy on climate change issues. The openness in informational policy also includes popularising scientific discoveries in this area through the mass media. Furthermore, the doctrine acknowledges the mass media as one of the actors in climate change regulation policy. According to this document – realising a possible clash of interests in climate change policy – the mass media will have to be socially responsible and enter the process of preventing conflicts and social tension, and lobbying of certain actors such as oil companies (Climate Doctrine 2009).

The adoption of the doctrine (just as any other document on climate change) caused heated debate in Russia's scientific and political community. When the doctrine was accepted, many saw it as a positive move in domestic climate change policy. The rector of the Russian State Hydrometeorological University, Professor Lev Karlin (2010), in an opinion piece for the website of the Russian Environmental NGO 'Bellona', said that it was definitely upbeat news, the fact that the state had turned towards the opinion that climate change problems would stimulate scientific research. However, a few years after it became obvious that the doctrine did not produce any practical outcomes, opinions on it became more pessimistic – 'the doctrine seems as irrelevant and abandoned now as a framework document can possibly be' (Dobrovidova 2012c).

In this sense, Russia's position as announced at the Copenhagen Conference can also be considered as a positive shift in its climate change policy. Alexey Kokorin from WWF-Russia stated (interview 14, Moscow, 27 July 2011), 'it is difficult to judge whether Russian climate change policy is good or bad. Probably by the European standard it is awful, but for Russia the fact that the anthropogenic character of climate change is admitted already means a lot'.

Other manifestations of the change in Russia's climate policy are represented by the approval of the Climate Doctrine Implementation Plan (2011) and the creation of the position of presidential adviser for climate change. Prior to the conference on 27 November 2009 Alexander Bedritsky was appointed the president's adviser on climate change and he himself noted that his appointment in this new position demonstrated that the importance of state climate policy was rising[18] (Bedritsky 2011). This change was reinforced in March 2010 at the meeting of Russia's Security Council[19] which focused 'on measures to prevent threats to national security in relation to global climate change'. Here, Medvedev stated that even though it was unclear what the prospects of international negotiations on the problems of climate change were, Russia, as a responsible state would follow its chosen strategy – the development of a sustainable economy and 'so-called green technologies' by creating a modern energy sector and reducing carbon emissions. Medvedev underlined that it was necessary to develop a strategy which would help to prevent or minimise climate change and would also preserve the country's economic competitiveness in its major export positions (President of Russia website 2010c).

In addition, it must be noted that unlike several years ago during the Kyoto ratification negotiations, Russian scientists became more unified in their opinion on the threat that climate change posed for Russia and the world. As Vladimir Berdin said (interview 4, Chemal, 13 August 2011), '[we scientists] absolutely agree with the IPCC conclusions and we have our own contributors to their reports, so at this moment the position of Russian scientists is identical to the majority of their colleagues around the world'. Perhaps a slight change in state policy will give scientists as well as other members of the 'epistemic community' on the issue of climate change a 'policy window'[20] (Evangelista 1995) in order to lobby their interests (see more on the role of scientists in Russian climate policy in Wilson Rowe 2013).

Some scholars connect this modification in the state's climate change policy with Medvedev's presidency and his policy of economic modernisation (Sharples 2013), including improvement in energy efficiency. Medvedev's personal view on the modernisation of Russia has been presented in his infamous article 'Go Russia!' [*Rossiya, Vpered!*] (Medvedev 2009), where he critiques 'a primitive economy based on raw materials' which is 'humiliating' for the modern state of Russia. Henry and Sundstrom (2012) argue that during Medvedev's presidency Russia's climate policy depended on his 'authority, the degree to which energy efficiency goals have been institutionalised, and the economic incentives and constraints of a post-Kyoto agreement that induces Russia to participate' (ibid, p. 1316). The following section looks at Medvedev's official speeches during his time in office which relate to climate change. It is argued that the official political discourse was not linked specifically to Medvedev's presidency, but rather reflects embedded ideas pertaining to the economic benefits that should remain even after the change at the executive level.

Presidential speeches on climate change: 'Either we all should contribute, or we should abandon all attempts'

There are a number of scholars who analyse policy speeches and official documents to study Russia's domestic and foreign policies (see e.g., Angermueller 2012, Kratochvil 2008, Kratochvil et al. 2006, O'Loughlin et al. 2004). Jensen and Skedsmo (2010, p. 441) in their comparative study of Russian and Norwegian Arctic policies justify their choice of data on the grounds that 'the selected texts are all articulated by formal political authority [and] intend to represent the countries' approaches to the European Arctic[. . . . The texts] set the agenda and shape the issues at hand, and they frame and produce representations of foreign policy.' For the purpose of this research, presidential speeches are treated as the written representations of the state leaders' approach to Russia's climate change policy.

The analysed texts were collected from the official website of the Russian president (http://kremlin.ru) and include publicly available transcripts of the president's statements at press conferences, interviews and meetings with government officials (foreign and domestic) and to the general public. The 72 speeches studied cover Medvedev's presidency (May 2008–May 2012) – all mention 'climate change'. The data were further analysed using qualitative content analysis. While Krippendorff (2004) questions the categorisation of content analysis as quantitative and qualitative, in this case the 'qualitative' attribute means that the relatively small number of texts were individually studied and specified keywords ('climate change') were analysed within the textual context. Further on, other content analysis methods are employed – 'taking a sample of media, establishing categories of content, measuring the presence of each category within a sample, and interpreting the result' (Bertrand and Hughes 2005, p. 198). As the content analysis here is 'problem-driven' (Krippendorff 2004), the defined textual categories are influenced by the research questions aimed at exploring Medvedev's approach to the problem of climate change and whether the change in Russia's climate policy could be solely ascribed to his presidency. Six categories were identified within the studied texts as follows: 'global cooperation', 'environmental leadership', 'economic benefits', 'helping the environment', 'global security' and 'responsibility' (several categories have been attributed to the same text); the results are presented in Table 3.2 below.

The least frequent category proved to be 'responsibility' (7 percent, $n = 5$) representing only a few cases of Medvedev acknowledging Russia's contribution to climate change and with only one case where Russia was specifically referred to as one of the biggest emitters in the world. This result, together with the relative unpopularity of the category 'helping the environment' (18 percent, $n = 13$), is quite predictable, based on the earlier discussion of the Russian government's neglect of environmental policy. The concept of 'environmental

Table 3.2 Percentage of Medvedev speeches (2008–2012) by identified categories

Category	%	Examples of quotations
Global cooperation	78	'the topic has not left anyone indifferent';
		'obviously, regardless of anyone's attitude, everyone should get involved based on scientific knowledge and objective predictions'
Economic benefits/ green economy	38	'we must improve our energy efficiency, which at the end will help to solve global problems of climate change and reduce the GHG emissions';
		'we should be ready for any scenario and use it for the benefit of our economy'
Global security	21	'another area of our cooperation is environmental security';
		'climate change is one of the main threats and challenges'
Helping the environment	18	'our goal is not only to improve our lives, but also to think about future generations; that is why the problem of climate change stays in the centre of our attention';
		'we all have an interest in radical improvement of our environment'
Environmental leadership	14	'currently Russia is a world leader of GHG emissions reduction';
		'some time ago we took very serious responsibilities, whilst a significant part of developing economies did not do it, such as China, India, Brazil; the Americans did not take action, but we did'
Responsibility	7	'we are all responsible for climate change';
		'we understand our responsibility for GHG emissions along with other major emitters'

leadership' also does not enter presidential discourse too often (14 percent, $n = 10$). Recurring throughout the speeches along with reminders of Russia's 'great commitments' to the Kyoto Protocol or 'drastic' goals of GHG emissions reduction is the more popular concept of presenting climate change as another 'global security' issue (21 percent, $n = 15$). This category acknowledges the importance of the climate change problem and stresses the urgent necessity to deal with it or adjust the state's policy, in view of climate change consequences. The latter message was largely provoked by the devastating consequences of the heat wave in Russia in the summer of 2010 – 'considering what is happening this summer, we do not know what is going to happen next year, the climate is changing, we have to take it into consideration and allocate some budget for it' (President of Russia website 2010a).

The top two results deserve more detailed discussion. Firstly, the majority of speeches (78 percent, $n = 56$) referred to climate change within the context of global cooperation. Also, 35 speeches from the category 'global cooperation' were presented during global summits (G8, G20, BRICS) or bilateral meetings with various state leaders where climate change was mentioned in the same sentence with other global challenges, such as global poverty, illegal immigration, energy, food security and so on. It could be argued that in the Russian case, climate change is used in order to demonstrate the state's involvement in processes of international cooperation and development as 'a modern and liberal state.' Another frequently repeated message in this category promotes the idea of global responsibility, where Medvedev appeals to every country to take part in the fight against climate change, as without global efforts solutions will not be found anyway. In this context, Russia's desire not to commit to the second period of the Kyoto process does not come as a surprise – it does not involve all countries. It should be noted that the majority of the speeches in the category 'global cooperation' mention climate change in only one or two sentences.

A more explicit discussion takes place within the 'economic benefits' category (38 percent, $n = 27$), in which the main message given by Medvedev is 'we will win no matter what', thus climate change will be addressed in the manner most beneficial for the country. For example, during the meeting with the managerial staff of Russia's Academy of Science just prior to the Copenhagen speech, Medvedev stated that 'development of an energy-efficient economy is a definite priority, regardless of our [Russia's] attitude towards climate change' (President of Russia website 2009b). A similar message was presented in Medvedev's official speech at the Copenhagen summit: a 'global climate "deal" is a real chance for "green" economic development and investments around the world. In the end, measures for mitigating climate change will assist in solving global environmental and socio-economic problems, in practice achieving those "millennium goals" we set some time ago' (President of Russia website 2009a). In his blog on 5 June 2010 (World Environment Day), Medvedev published a piece with the title 'Environment and Economy Do Not Contradict Each Other. A Normal Economy Is Environmentally Friendly.' In this article, Medvedev states that 'unfortunately with some delay, we have finally realised that it is vital to protect our environment and that economic and environmental developments are inextricably linked.' Further on in the article he once again talks about 'energy efficiency' and 'green economy' and how these ideas have become a trend which he finds quite sensible – 'I have always said that people start dealing with environmental problems when they feel the economic necessity.' At the meeting with the state security council 'on questions of environmental protection' (President of Russia website 2010c), Medvedev claimed that countries such as the United States and China got involved in climate change mitigation (and in general, problems of environmental development) because they saw the 'opportunity to make money, and we [Russia] should have the same attitude.'

The analysis of the presidential speeches mentioning 'climate change' suggests that whilst the categories such as 'environmental leadership' and 'responsibility' do not enter the official discourse that often, the category of 'economic benefits' proposes the most elaborated and explicit vision of climate change problems within the state's national interests, and the most frequent referral to the climate topic happens within the category of 'global cooperation.' It can be argued that Medvedev's presidency signified a shift from the policy of 'de-environmentalism' to *pragmatic environmentalism*, which holds that the environment will eventually benefit from the state's actions but only if it brings obvious benefits for Russia's state policies and/or economy. 'Without sensible pragmatism, we won't solve environmental problems.'[21]

Several years ago, Russia's environmental 'greatness' was just a 'cover' in the speeches of the country's leaders and state officials. In business terminology, this was what is called a 'green-washing technique' – 'tactics that mislead consumers regarding the environmental practices of a company' (Mason and Mason 2012 citing Parguel et al. 2011, p. 15). The 'consumers' here were the international community in front of which Russia tried to demonstrate its importance in the climate negotiation process, as seen in Russia's involvement in the Kyoto negotiations. Fear of possible economic losses due to international obligations to cut GHG emissions along with the underestimation of the negative effects of climate change resulted in Russia's peculiar involvement in the international negotiation processes and reluctant domestic climate change policy. Korppoo and Vatansever (2012) state that this (largely) superficial 'environmental leadership' does not work anymore. Firstly, the international community has realised that the reduction in GHG emissions was not a result of governmental policy; secondly, Russia (as the successor to the USSR) has significant historical responsibility for the world's GHG emission record; and thirdly, the Russian economy is still extremely carbon intensive (the carbon intensity of its GDP is 81 percent greater than the world average).

Recently, as the analysis of Medvedev's official speeches shows, there is a move to an understanding of climate change policy not as a policy of 'costs' but as one of 'opportunities' (Giddens 2010). As the director of the Centre of Environmental Policy, Vladimir Zakharov summarised, 'for the next 20 years nobody will be able to operate their economies without fossil fuels, so nothing threatens Russia's economic interests. Now we need to start thinking about how we can provide environmental services and get investments in the "greening" of our economy' (interview 29, Moscow, 21 July 2011).

Beside the involuntary drop in GHG emissions of the 1990s and a rather fictitious environmental leadership in climate change mitigation policy, Russia possesses the ability to lead the way in sustainable development without significant economic costs. Russia does have potential to de-carbonise its economy through an increase in energy efficiency and the development of renewable energy sources (RES) (Bagirov and Safonov 2010, Overland and Kjarnet 2009), which might benefit both Russia's economic development and global GHG reduction goals. Averchenkov (2009) argues that merely following its plans for the improvement of energy efficiency might be enough for Russia to fulfil its carbon reduction obligation by reducing GHG emissions by 40 percent (2007 is taken as a baseline) by 2050. There are a number of sectors which have potential for energy saving – for example, the municipal/utility sector (through the modernisation of central heating systems), the oil and gas sector (reduction of gas flaring or a decrease in leakage during gas transportation), transport (renovation and popularisation of public transport or implementation of fuel efficiency standards), residential buildings (through enforcing energy standards in new or renovated buildings as well as raising public awareness about energy saving and promoting the use of electricity meters) and so on (Averchenkov 2009, Opitz 2007, World Bank 2008). The extensive list of measures for developing the energy efficiency of the

Russian economy were formalised in Federal Law (23/11/2009, N261). The law was updated throughout the past several years, with the latest version, signed by the re-elected President Putin, extending measures for improving energy saving in the automobile sector.

Energy efficiency plans are already included in the Russian Energy Strategy Towards 2020 (Ministry of Energy of the Russian Federation 2003) and are considered to be a priority not only for the energy sector in particular but for the whole economy as well (Bogdan et al. 2009). In this sense, it is interesting how even the role of Gazprom in Russia's state policy can be presented as a tool for economic and environmental development. At the ceremony celebrating the start of building the offshore pipeline of the Nord Stream gas project, Medvedev stated that ' "Nord Stream" is not just a major transnational project, but also in our view, [Russia's] input into the global solution of environmental and climate problems [. . .] which will allow us to reduce GHG emissions without economic sacrifice' (President of Russia website 2010b).

With regard to renewable energy, the deputy director of the Russian State Institute of Energy Strategy, Pavel Bezrukikh, states that 'renewable energy sources in Russia could cover 35 percent of the country's total primary energy supply. [. . . Currently] renewable energy sources account for less than one percent of Russia's energy' (cited in Overland and Kjarnet 2009, p. 7). Overland and Kjarnet (2009, p. 5) state that the development of the capacity of Russia's renewable energy sector will allow the country to cut GHG emissions without any economic losses. The advance of RES is already included in Russia's Energy Strategy Towards 2020, which sees RES as a tool to ensure domestic energy security (by contributing to more stable supply of energy to the remote regions of the Far North of Siberia) as well as a solution to environmental problems. The strategy aims to increase the share of renewable energy sources in Russia by up to 4.5 percent by 2020. The development of energy efficiency and renewable energy sector was also formalised through the acceptance of the following official documents – a Presidential Decree 'On some measures to improve energy and environmental performance of the Russian economy' (2008), a resolution of the Russian government '[o]n the main directions for the state policy to improve the energy efficiency of the electricity sector on the basis of renewable energy sources for the period up to 2020' (2009) and the earlier mentioned Federal Law 'on energy saving and energy efficiency'.

According to a report by the World Bank (2008), these savings through improvements in energy efficiency[22] will benefit the economy by approximately $120–150 billion per annum through an increase in oil and gas exports. Russia will also decrease its GHG emissions, improve its air quality and as a consequence lessen the health risk from pollution for the population. A report from the McKinsey Global Institute (2009, p. 7) states that Russia 'has the largest relative potential among all the BRIC[S] countries to reduce emissions through implementing only measures that are economically attractive'. The McKinsey Global Institute proposes 60 measures which would require investments of €150 billion (over twenty years), but the energy savings achieved would result in €345 billion (over the same timeframe), energy consumption would be reduced by 23 percent and GHG emissions by 19 percent.

Another issue connected with the economics of climate change should also be highlighted: Russia's economic losses due to the consequences of climate change. Arguably, this idea was implicitly or in some cases explicitly present in Medvedev's speeches in the category of 'global security', in which climate change is treated as another major threat to national security. Indeed, as was mentioned earlier in this chapter, Russia is extremely vulnerable to the consequences of climate change, including threats to the economic stability of the country. According to data provided by Roshydromet (2012), every year Russia's economy loses 60 billion rubles (around £1.27 billion) due to extreme weather events and climate change; this amount increases every year by 6 percent. For instance, in summer 2010 the central part of Russia saw continuous records for highest temperature and was subject to massive areas of toxic smog. The heat led to significant damage to the agricultural sector, resulting in losses of 41.6 billion rubles (around £0.832 billion) (RIA Novosti 2010a). The negative consequences of the anomalous weather event also led to severe social losses: during these months the death rate in Moscow alone increased from 360–380 people dying a day to 700 (RIA Novosti 2010b). A report by Roshydromet (2014) confirmed that whilst the heatwave cannot be directly connected with anthropogenic climate change, the increasing frequency and intensity of this type of weather event is stimulated by human-induced global warming. As an environmental activist (interview 27, Moscow, 22 July 2011) stated, 'even though the connection to climate change was not proven, because the fires [stimulated by the heatwave] happened in Moscow, it forced our government to doubt their position towards climate change'. Another catastrophic example of an anomalous weather event were floods in summer 2012 in Krasnodar krai, which resulted in numerous deaths (especially in the town of Krymsk) (Roshydromet 2014). According to Climate Change Risk Index (Kreft and Eckstein 2013), Krasnodar's floods caused $400 million in overall damage, which subsequently moved Russia to 9th position in the global Climate Change Index in 2012 (whilst in 2011 it occupied the 95th line of the report and overall in the twenty-year period [1993–2012], it was considered the 27th most vulnerable country).

As the above analysis has demonstrated, for Russia, climate change mitigation policy has become increasingly beneficial in both environmental and (perhaps more importantly for the state) economic terms. Commitment to the GHG emissions reduction goals is a 'low hanging fruit', where by modernising its economy and getting more income into the budget, Russia can also contribute to the global fight against climate change and become a more genuine 'environmental leader'. Moreover, there is also a realisation of the economic losses which Russia is already facing due to the consequences of climate change, which is arguably also pushing the issue up the priority ladder. As the content analysis of Medvedev's speeches has showed, Russian state leaders become more and more aware of the economic side of climate policy.

Concluding remarks

On 1 January 2013 the second commitment period of the Kyoto Protocol came into force. Russia became one of the few Annex I parties (along with Japan and New Zealand) that refused to take on new targets in GHG emissions reduction within this international framework (RIA Novosti 2012c), which was announced at the UNFCCC in Doha, Qatar (2012). Two months earlier, the Russian president's adviser on climate change, Bedritsky, justified the state's position by stating that Russia advocated the adequate involvement of all countries without any exception in finding solutions to the problem of climate change (RIA Novosti 2012a; see more in *International Affairs* 2010, p. 237).

This recent Russian participation in the international climate change negotiations in Qatar should not come as a surprise or be seen as a totally negative development. First, Russia did not 'go Canadian' (Dobrovidova 2012a), meaning that unlike Canada, Russia did not completely abandon the protocol. At least until 2015 it will be (along with other parties) calculating the results of GHG emissions reductions during the first period (2008–2012), and even after that it will keep reporting its emissions levels according to the protocol's requirements (RIA Novosti 2012c). Since the Copenhagen Conference, the rhetoric used by Russian officials has changed: even though they remain reserved towards international commitments, all recent statements confirm domestic dedication towards GHG emissions reduction, which will be achieved through economic modernisation and improvements in energy efficiency. The recent changes in climate change policy emerged because of the realisation that Russia can develop its economy and cut GHG emissions (what is summarised here as 'climate pragmatism'). This again leads us to the deep connection between the economy and the environment in Russia. Under new circumstances this correlation can actually be seen as a positive tendency. Practically speaking, the idea of treating climate change mitigation as profitable and beneficial as well as understanding the nature of economic losses from climate degradation and the importance of climate change as a topic of global concern is moving the policy beyond the point of rhetoric. This is encouraging Russia to take steps towards a more sustainable and 'greener' economy and consequently making Russia a more realistic 'environmental leader'. If the hypothesis that the media coverage of climate change issues correlates with the state's policy is correct, then it will be possible to observe this change in the state's policy on the pages of newspapers.

Notes

1 Tsalikov (2009) stresses that the biggest danger of melting permafrost is in the region of Novaya Zemlya – an area of nuclear waste storage.
2 For instance, in the Arkhangelsk region, the necessity to import coal and oil for its industrial needs from other regions has led to a great interest in cooperating with the 'energy saving and environmental investment agencies in order to improve the attractiveness of implementing the Kyoto mechanisms' (Buchner and Dall'Olio 2005, p. 363).
3 JI mechanisms (or projects) allow a country with an emissions reduction or limitation commitment under the Kyoto Protocol (Annex B Party) to earn emissions reduction

units from an emission-reduction or emission-removal project in another Annex B Party, which can be counted towards meeting its Kyoto target (UNFCCC 2011).

4 As an example, one might think of the 'grand' Soviet idea of the Siberian river reversal, when instead of allowing northern rivers to 'uselessly' fall into the Arctic Ocean, Soviet scientists came up with a plan of diverting them towards the densely populated and agriculturally valuable territories of Central Asia.

5 Yanitsky (2009) based his conclusions on the extensive field research he conducted over two decades (1985–2007), his personal experience of working in UNESCO's 'Man and the Biosphere' programme, as well as participation in a number of international research projects.

6 After the dissolution of Goskomekologiya, Jo Crotty (2003) conducted research in one of Russia's regions – Samara Oblast – and she argued that at the regional level the institutional restructuring was not that noticeable; 'the monitoring and control function of the old environmental bureaucracy had been largely retained, albeit with some staff cuts, under the new Ministry' (ibid, p. 473). In a more recent article, Crotty and Rodgers (2009) reinvestigate the case of Samara Oblast and conclude that after the merger of Goskomekologiya and the Ministry of Natural Resources, 'there has not been a subsequent decline but in fact an expansion of bureaucratic controls and regulatory bodies in the area of environmental protection in Russia' (ibid, p. 12).

7 Roshydromet is a federal executive authority that provides public services in hydro-meteorology, environmental monitoring, pollution, observation of the influence on the meteorological and other geophysical processes. It ensures that Russia fulfils obligations under international treaties, including the Convention of the World Meteorological Organization, the UNFCCC and the Protocol on Environmental Protection to the Antarctic Treaty (Roshydromet 2011).

8 According to the EIA (2010), these reserves contain 1,680 trillion cubic feet, representing about 25 percent of the world's reserves. Most of these reserves are in Siberia. Russia is not only one of the world's largest gas producers, but it is also the biggest exporter of gas in the world. Oil reserves are also the second largest in the world, making up 60 billion barrels, which again are mostly situated in Siberia (specifically the western part).

9 One of the reasons for this is the severe weather conditions in which most of the industries are situated, making the production process more energy intensive (Perelet et al. 2007, Shaw 2009). Another reason is that along with being a leader in the gas and oil industry comes leadership in the amount of gas flaring. This side effect of oil production is responsible for 84,000 tons of GHG emissions a year (Cnews.ru 2007).

10 King (2012) states that according to the 2008 UN report, 'Russia was the most generous country in the world when it comes to fossil fuel subsidies, spending $40 billion annually to support those industries'.

11 Amongst them are Gazprom, LUKoil, Norilsk Nickel, UES Rossiya, joint-stock company 'Rusal' and Arkhangelsk Pulp and Paper Mill (Bogdan et al. 2009).

12 Sberbank (the Savings Bank of the RF, which was authorised by the Russian government to approve and select JI projects) estimated that Russian projects have a potential to cut 1.2–1.5 billion tons of CO_2 and bring into the country direct carbon investments of 250–300 billion rubles (around £50–60 billion) by 2020 (Men'she Dvukx Gradusov 2012).

13 Oldfield (2005, p. 81, citing Missfeldt and Villavicenco 2000, p. 382) suggests that the GHG emissions drop would have been even more significant at that time if Russia's energy efficiency had not worsened.

14 As was unofficially stated amongst the people involved in Russia's climate change affairs, it was almost certain that Illarionov's work and position was funded by international fossil fuel companies (interview 31, Moscow, 27 July 2011), in this sense the role of this scientist could be compared to the role of the conservative movement in the United States in advocating the climate sceptic position (McCright and Dunlap 2003).

15 During the Kyoto negotiations, 250 representatives of Russian science signed a document in support of the protocol, and in the media more and more statements from

Russian academics could be found seeking to change the Russian people's and government's attitude towards the problem.

16 Joining the WTO was one of Russia's key policy goals. Throughout the 1990s it made several attempts to negotiate its entrance into the organisation; however, it kept failing to do so. One of the EU's demands was for Russia to even out its gas prices between its internal and external markets. In May 2004, during the EU-Russia summit, agreement was reached that Russia would liberalise its banking and telecommunications sectors, decrease its import tariffs, even out its gas prices by 2010 and ratify the Kyoto Protocol (Buchner and Dall'Olio 2005).

17 The Fifteenth Conference of the Parties under the UNFCCC (COP-15) was held from 7 December to 18 December 2009 in Copenhagen, Denmark. Taking into account that the Kyoto Protocol was due to expire at the end of 2012 and that a new document needed to be introduced, many politicians, scientists and NGOs set big hopes on the Copenhagen Conference, expecting that countries would be able to compromise and reach some degree of agreement. However, even before the conference started, it was already apparent that it was very likely that agreement would not be achieved. When the conference was over, the word 'failure' was commonly used to describe it. The main problem was that the conference did not manage to produce any legally binding document with positions similar to those of the Kyoto Protocol. Among the major problems were disagreements between industrialised and developing countries as well as within the camp of the industrialised countries. On the other hand, the Copenhagen Conference could not be called a complete fiasco. The Copenhagen Accord, while not with the legal power of the Kyoto Protocol, signified a level of agreement among more than 25 countries. The accord concluded that the world's goal is to keep temperature rises under 2°C (Bodansky 2010) and it allocated the budget for the mitigation and adaptation processes. But it did not set concrete emissions reduction targets; the 2°C temperature limit is quite questionable and has a more political than scientific basis; and the accord does not place countries under an obligation but only recommends sticking to its positions.

18 Referring again to the work of Peter Haas (1989) on the role of epistemic community, the appointment of Bedritsky could be paralleled with the penetration of marine scientists in the Algerian government and the consequent growth of their influence over the state's marine environmental regime.

19 Russia's Security Council draws up the major official documents on Russia's national and international policy where security threats are evident (Presidential Decree on the Security Council 2011).

20 Looking at the role of transnational actors in the security policy of the Soviet Union in the 1980s, Matthew Evangelista (1995) comes to the conclusion that even though the majority of these transnational actors were scientists with an extensive level of expertise and competence on the issue, their opinion was only taken into account when the Soviet domestic structure was shaken due to the 'severity of the economic crisis, the challenges of the Reagan administration, and the advent of a strong reformist leader' (ibid, p. 36).

21 This statement comes from Medvedev's speech on climate change (President of Russia website 2010c).

22 The report states that 'Russia's current energy inefficiency is equal to the annual primary energy consumption of France' (World Bank 2008, p. 5).

Bibliography

Afionis, S. and Chatzopoulos, I. (2010) 'Russia's Role in UNFCCC Negotiations since the Exit of the United States in 2001', *International Environmental Agreements – Politics, Law and Economics*, 10: 45–63.

Aleksashenko, S. (2012) 'Russia's Economic Agenda', *International Affairs*, 88/1: 31–48.

Andonova, L. (2008) 'The Climate Regime and Domestic Politics: The Case of Russia', *Cambridge Review of International Affairs*, 21: 483–504.

Angermueller, J. (2012) 'Fixing Meaning. The Many Voices of the Post-Liberal Hegemony in Russia', *Journal of Language and Politics*, 11/1: 115–134.

Averchenkov, A. (2009) *Ekonomika i Klimat: Uchastie Rossii v Reshenii Global'noy Ekologicheskoy Problemy*, Moscow: Institut Ustoychivogo Razvitiya/Tsentr Ekologicheskoy Politiki Rossii.

Bagirov, A. and Safonov, G. (2010) *Energobezopasnost' i Klimat: Global'nye Vyzovy dlya Rossii*, Moscow: Teis.

Bedritsky, A. (2011) 'Izmenenie Klimata: Prioritety Deystviy i Grazhdanskoe Obshchestvo', *Na Puti k Ustoychivomu Razvitiyu Rossii*, 55: 11–15.

Bertrand, I. and Hughes, P. (2005) *Media Research Methods. Audiences, Institutions, Texts*, Basingstoke: Palgrave Macmillan.

Bodansky, D. (2010) 'The International Climate Change Regime: The Road from Copenhagen', *Viewpoints Series, Harvard Project on International Climate Agreements*, http://papers.ssrn.com/sol3/papers.cfm?abstract_id=1693889, date accessed 20/11/2010.

Bogdan, L., Dobrolyubova, Y., Kozel'tsev, M. and Surovikina, E. (2009) *Politika i Deyatel'nost' Rossii v Oblasti Ekologii i Izmeneniya Klimata*, Rossiyskiy Regional'nyy Ekologicheskiy Tsentr: Moscow.

Buchner, B. and Dall'Olio, S. (2005) 'Russia and the Kyoto Protocol: The Long Road to Ratification', *Transition Studies Review*, 12: 349–382.

Charap, S. (2010) 'Russia's Lacklustre Record on Climate Change', *Russian Analytical Digest*, 79: 11–15.

Climate Change Action Plan Report (1999), www.gcrio.org/CSP/pdf/russianfed_snap.pdf, date accessed 11/12/2009.

Climate Doctrine of the Russian Federation (2009), http://archive.kremlin.ru/eng/text/docs/2009/12/223509.shtml, date accessed 10/01/2010.

Cnews.ru (2007) 'Rossiya Lidiruet po Ob'emy Fakel'nogo Szhiganiya Gaza', www.cnews.ru/news/line/index.shtml?2007/08/31/264472, date accessed 15/03/2011.

Comprehensive Implementation Plan of the Climate Doctrine of the Russian Federation for the Period up to 2020, 25 April 2011. Directive No. 730-p of the Government of the Russian Federation.

Constitution of the Russian Federation (1993), www.constitution.ru/en/10003000-01.htm, date accessed 2/4/2015.

Crotty, J. (2003) 'The Reorganization of Russia's Environmental Bureaucracy: Regional Response to Federal Change', *Eurasian Geography and Economics*, 44/6: 462–475.

Crotty, J. and Rodgers, P. (2009) 'The Re-organisation of Russia's Environmental Bureaucracy – Good News or Bad?'; paper presented at the Economic and Social Research Council Sustainable Development Conference, Utrecht, Holland (1 September).

Dobrovidova, O. (2012a) 'Comment: Think You Know Russia's Position on Climate Change? Think Again', *Responding to Climate Change* (RTCC) (19 October), www.rtcc.org/comment-think-you-know-russias-position-on-climate-change-think-again/, date accessed 24/10/2012.

Dobrovidova, O. (2012b) 'Comment: Why Russia's Climate Change Policy Is Like a Cargo Cult?' *RTCC* (21 September), www.rtcc.org/policy/comment-why-russias-climate-change-policy-is-like-a-cargo-cult/, date accessed 15/10/2012.

Dobrovidova, O. (2012c) 'Comment: Why Russia's Position on the Kyoto Protocol Is a Mystery Even to Its Own Ministers', *RTCC* (27 November), www.rtcc.org/comment-why-russias-position-on-the-kyoto-protocol-is-a-mystery-even-to-its-own-ministers/, date accessed 30/11/2012.

EIA (2010) *Country's Analysis Briefs: Russia*, www.eia.gov/countries/cab.cfm?fips=RS, date accessed 2/4/2015.

Ekoreporter (2012) 'Rossiya Poluchila Zvanie "Iskopaemoe Dnya"', http://ecoreporter.ru/node/1312, date accessed 30/11/2012.

Evangelista, M. (1995) 'The Paradox of State Strength: Transnational Relations, Domestic Structures, and Security Policy in Russia and the Soviet Union', *International Organisation*, 49/1: 1–38.

Fay, M., Block, R. and Ebinger, J. (eds.) (2010) *Adapting to Climate Change in Eastern Europe and Central Asia.* Washington, DC: The World Bank.

Feldman, D. and Blokov, I. (2009) 'Promoting an Environmental Civil Society: Politics, Policy, and Russia's Post-1991 Experience', *Review of Policy Research*, 26: 729–759.

Feldman, D. and Blokov, I. (2012) *The Politics of Environmental Policy in Russia*, Cheltenham: Edward Elgar.

Firsova, A. and Taplin, R. (2009) 'Australia and Russia: How Do Their Environmental Policy Processes Differ?' *Environment, Development and Sustainability*, 11/2: 407–426.

Gazprom (2014) 'Kyoto Protocol', www.gazprom.com/nature/kioto/, date accessed 09/09/2014.

Giddens, A. (2010) 'Russia Can Ill-Afford Climate Cavalierism', *Policy Network* (31 August), www.policy-network.net/pno_detail.aspx?ID=3884&title=Russia-can-ill-afford-climate-cavalierism, date accessed 13/12/2010.

Gotz, R. (2007) 'Russian and Global Warming – Implications for the Energy Industry', *Russian Analytical Digest*, 23: 11–13.

Government of the Russian Federation (23 November 2009). Federal Law N261-FZ 'On Energy Saving and Energy Efficiency'.

Government of the Russian Federation (8 January 2009). Resolution no. 1-r '[O]n the Main Directions for the State Policy to Improve the Energy Efficiency of the Electricity Sector on the Basis of Renewable Energy Sources for the Period up to 2020'.

Haas, P. (1989) 'Do Regimes Matter? Epistemic Communities and Mediterranean Pollution Control', *International Organisation*, 43/3: 377–403.

Henry, L. (2010) 'Between Transnationalism and State Power: the Development of Russia's Post-Soviet Environmental Movement. *Environmental Politics*, 19, 756–781.

Henry, L. and Douhovnikoff, V. (2008) 'Environmental Issues in Russia', *Annual Review of Environment and Resources*, 33: 437–460.

Henry, L. and Sundstrom, L. (2007) 'Russia and the Kyoto Protocol: Seeking an Alignment of Interests and Image', *Global Environmental Politics*, 7: 47–69.

Henry, L. and Sundstrom, L. (2012) 'Russia's Climate Policy: International Bargaining and Domestic Modernisation', *Europe-Asia Studies*, 64/7: 1297–1322.

Illarionov, A. and Pivovarova, N. (2004) 'Ekonomicheskie Posledstviya Ratifikatsii Rossiyskoy Federatsiey Kiotskogo Protokola', *Voprosi Ekonomiki*, 11.

International Affairs (2010) 'Russia and the World after Copenhagen', 56/6: 235–258.

Izvestiya (1997) 'Rossiya Mozhet Zarabotat' na Poteplenii Klimata', 16 December.

Jensen, L. and Skedsmo, P. (2010) 'Approaching the North: Norwegian and Russian Foreign Policy Discourses on the European Arctic', *Polar Research*, 29/3: 439–450.

Karlin, L. (2010) 'Klimaticheskaya Doktrina: Blago ili Blazh?' www.bellona.ru/articles_ru/articles_2010/climate-doctrine-blago, date accessed 10/02/2011.

Kattsov, V. and Porfirev, B. (eds.) (2011) *Otsenka Makroekonomicheskix Posledstviy Izmeneniy Klimata na Territorii Rossiyskoy Federatsii na Period do 2030 Goda i Dalneyshuyu Perspektivu*, Moscow: Rosgidromet.

Kattsov, V., Meleshko, V. and Chicherin, S. (2007) 'Izmenenie Klimata i Natsional'naya Bezopasnost' Rossiyskoy Federatsii', *Pravo i Bezopasnost'*, 1-2/22–23, http://dpr.ru/pravo/pravo_20_5.htm, date accessed 13/02/2011.

King, E. (2012) 'What's Stopping Russia Taking a Lead at the UN Climate Change Talks?' *RTCC* (13 August), www.rtcc.org/whats-stopping-russia-taking-a-lead-at-the-un-climate-change-talks/, date accessed 24/10/2012.

Klyuev, N. (2002) 'Rossiya: Ekologicheskiy "Portret"', *Ekologiya i Zhizn'*, 6/29: 16–19.

Klyuev, N. (2007) 'Natsional'nye Ekologo-Promyshlennye Piramidy', *Ekologiya i Zhizn'*, 11/72: 18–21.

Korppoo, A. (2008) 'Russia and the Post-2012 Climate Regime: Foreign Rather than Environmental Policy', *UPI Briefing Paper*, 23.

Korppoo, A. and Vatansever, A. (2012) *Klimaticheskoe Videnie dlya Rossii: Ot Ritoriki k Deystviyu*, Moscow: Carnegie Endowment for International Peace.

Kotov, V. (2002) 'Policy in Transition: New Framework for Russia's Climate Policy', www.feem.it/web/activ/_activ.html, date accessed 16/01/2012.

Kratochvil, P. (2008) 'The Discursive Resistance to EU-enticement: The Russian Elite and (the Lack of) Europeanisation', *Europe-Asia Studies*, 60/3: 397–422.

Kratochvil, P., Cibulkova, P. and Benes, V. (2006) 'Foreign Policy, Rhetorical Action and the Idea of Otherness: The Czech Republic and Russia', *Communist and Post-Communist Studies*, 39/4: 497–511.

Kreft, S. and Eckstein, D. (2013) *Global Climate Risk Index 2014: Who Suffers Most from Extreme Weather Events? Weather-Related Loss Events in 2012 and 1993 to 2012. Briefing Paper*, https://germanwatch.org/en/download/8551.pdf, date accessed 08/09/2014.

Krippendorff, K. (2004) *Content Analysis: An Introduction to Its Methodology*, London: Sage.

Kuraev, S. (2011) 'K Voprosy o Formirovanii Klimaticheskoy Politiki v Rossii', *Russian Regional Environmental Centre*, www.rusrec.ru/en/node/167, date accessed 10/02/2011.

Kyoto Protocol to the United Nations Framework Convention on Climate Change (1998), http://unfccc.int/resource/docs/convkp/kpeng.pdf, date accessed 2/4/2015.

Mandrillon, M.-H. (2008) 'Debating Kyoto: Soviet Networks and New Perplexities'. In: S. White (ed.) *Media, Culture and Society in Putin's Russia*, Basingstoke: Palgrave Macmillan: 133–153.

Mason, M. and Mason, R. (2012) 'Communicating a Green Corporate Perspective: Ideological Persuasion in the Corporate Environmental Report', *Journal of Business and Technical Communication*, 26/4: 479–506.

Mason, R. (2011) 'Gazprom's Gas Market Dominance Threatened: Russian Giant Facing a Challenge to Its Role as the Best Exporter of Gas for Europe', *Daily Telegraph* 22 February: 5.

McCright, A. and Dunlap, R. (2003) 'Defeating Kyoto: The Conservative Movement's Impact on US Climate Change Policy', *Social Problems*, 50/3: 348–373.

McKinsey Global Institute (2009) 'Lean Russia. Sustaining Economic Growth through Improved Productivity', www.mckinsey.com/insights/winning_in_emerging_markets/lean_russia_sustaining_economic_growth, date accessed 7/03/2013.

Medvedev, D. (2009) 'Rossiya, Vpered!', www.kremlin.ru/news/5413, date accessed 09/09/2014.

Medvedev, D. (2012) 'Rost Ekonomiki i Sberezhenie Prirody – Strategicheskie Zadachi, i Oni Dolzhny Byt' Sbalansirovany', http://blog.da-medvedev.ru/post/235/transcript, date accessed 10/02/2013.

Men'she Dvukh Gradusov (2012) 'Kiotskiy Protokol Prines Rossii Bolee Milliarda Dollarovykh Investitsiy' (18 November), http://below2.ru/2012/11/18/kprofit/, date accessed 4/12/12.

Ministry of Energy of the Russian Federation (2003) 'The Summary of the Energy Strategy of Russia for the Period of Up to 2020', http://ec.europa.eu/energy/russia/events/doc/2003_strategy_2020_en.pdf, date accessed 10/07/2012.

Oldfield, J. (2001) 'Russia, Systemic Transformation and the Concept of Sustainable Development', *Environmental Politics*, 10: 94–110.

Oldfield, J. (2005) *Russian Nature. Exploring the Environmental Consequences of Societal Change*, Hants: Ashgate.

Oldfield, J. and Shaw, D. (2002) 'Revisiting Sustainable Development: Russian Cultural and Scientific Traditions and the Concept of Sustainable Development', *Area*, 34/4: 391–400.

Oldfield, J., Kouzmina, A. and Shaw, D. (2003) 'Russia's Involvement in the International Environmental Process: A Research Report', *Eurasian Geography and Economics*, 44/2: 157–168.

O'Loughlin, J., O Tuathail, G. and Kolossov, V. (2004) 'Russian Geopolitical Storylines and Public Opinion in the Wake of 9-11: A Critical Geopolitical Analysis and National Survey', *Communist and Post-Communist Studies*, 37/3: 281–318.

Opitz, P. (2007) 'Energy Savings in Russia – Political Challenges and Economic Potential', *Russian Analytical Digest*, 23: 5–7.

Orlenok, V., Kurkov, A., Kucheryavyy, P. and Tupikin, S. (1998). *Fizicheskaya Geografiya*, Kaliningrad: KGY.

Overland, I. and Kjarnet, H. (2009) *Russian Renewable Energy. The Potential for International Cooperation*, Surrey: Ashgate.

Parker, L. and Blodgett, J. (2008) *Greenhouse Gas Emissions: Perspectives on the Top 20 Emitters and Developed versus Developing Nations*, US Congressional Research Service, Report for Congress, RL32721.

Perelet, R., Pegov, S. and Yulkin, M. (2007) 'Climate Change. Russia Country Paper', *Human Development Report* 2007/2008.

Peterson, D. J. and Bielke, E. (2001) 'The Reorganization of Russia's Environmental Bureaucracy: Implications and Prospects', *Post-Soviet Geography and Economics*, 42/1: 65–76.

Porfiriev, B. (1997) 'Environmental Policy in Russia: Economic, Legal and Oganisational Issues', *Environmental Management*, 21/2: 147–157.

President of the Russian Federation (2008, 4 June) Decree no. 889 'On Some Measures to Improve Energy and Environmental Performance of the Russian Economy.'

President of Russia website (2009a) 'Konferentsiya OON po Problemam Global'nogo Izmeneniya Klimata', http://kremlin.ru/news/6384, date accessed 15/03/2011.

President of Russia website (2009b) 'Stenograficheskiy Otchet o Vstreche s Rukovodstvom Rossiyskoy Akademii Nauk', http://xn-d1abbgf6aiiy.xn-p1ai/transcripts/6344, date accessed 18/02/2013.

President of Russia website (2010a) 'Po Porucheniyu Prezidenta Budet Razrabotana Programma Pereosnashcheniya Pozharnoy Okhrany', www.kremlin.ru/news/8563, date accessed 18/02/2013.

President of Russia website (2010b) 'Vystuplenie na Tseremonii Nachala Stroitel'stva Morskoy Chasti Gazoprovoda "Severnyy Potok"', http://kremlin.ru/transcripts/7410, date accessed 20/02/2013.

President of Russia website (2010c) 'Zasedanie Soveta Bezopasnosti po Voprosam Izmeneniya Klimata', www.kremlin.ru/news/7125, date accessed 1/03/2011.

Presidential Decree on the Security Council (2011), www.scrf.gov.ru, date accessed 3/03/2013.

RIA Novosti (2010a) 'Posledstviya Anomal'no Zharkogo Leta 2010 Goda', http://ria.ru/eco/20101126/301092668.html, date accessed 11/04/2013.

RIA Novosti (2010b) 'Deadly Moscow Summer Kills 700 Daily' (9 August), http://en.rian.ru/russia/20100809/160126534.html, date accessed 16/03/2011.

RIA Novosti (2012a) 'Medvedev Vernul 'Kiotskiy' Vopros v Povestku Dnya' (18 October), www.ria.ru/eco/20121018/904127750-print.html, date accessed 24/10/2012.

RIA Novosti (2012b) 'Veroyatnost' Smeny Pozitsii RF po Kioto-2 Blizka k Nulyu – Bedritsky' (21 November), http://ria.ru/science/20121121/911644186-print.html, date accessed 30/11/2012.

RIA Novosti (2012c) 'Vybor RIA Novosti: Glavnye Ozhidaemye Sobytiya 2013 Goda v Ekologii', http://ria.ru/eco/20121225/916135021.html, date accessed 25/01/2013.

Roshydromet (2008) *Otsenochnyy Doklad ob Izmeneniyakh Klimata i Ikh Posledstviya na Territorii Rossiyskoy Federatsii. Obshchee Rezyume*, Moscow: Rosgidromet, http://climate2008.igce.ru/v2008/pdf/resume_ob.pdf, date accessed 15/02/2011.

Roshydromet (2011) *O Nashey Slyzhbe*, www.meteorf.ru/rgm2d.aspx, date accessed 6/05/2011.

Roshydromet (2012) *Izmeneniya Klimata*, 30.

Roshydromet (2014) *Second Assessment Report of Climate Change and Its Consequences in the Russian Federation*. Moscow: Roshydromet.

RSEU [Russian Social-Ecological Union] (2012) 'Rossiya i Klimaticheskoe Anti-Liderstvo', www.rusecounion.ru/klimat_doha_221212, date accessed 20/02/2013.

Sharples, J. (2013) 'Russian Approaches to Energy Security and Climate Change: Russian Gas Exports to the EU', *Environmental Politics*, 22/4: 683–700.

Shaw, D. (2009) 'Russia: A Geographic Preface'. In: M. L. Bressler (ed.) *Understanding Contemporary Russia*, Boulder, CO: Lynne Rienner Publishers: 7–32.

Tipton, J. (2008) 'Why Did Russia Ratify the Kyoto Protocol? Why the Wait? An Analysis of the Environmental, Economic, and Political Debates', *Slovo*, 20: 67–96.

Tsalikov, R. (2009) 'Izmeneniya Klimata na Severe Rossii: Opasnosti i Ugrozy Zhiznedeyatel'nosti', *Region: Ekonomika i Sotsiologiya*, 1: 158–166.

Tynkkynen, N. (2010) 'A Great Ecological Power in Global Climate Policy? Framing Climate Change as a Policy Problem in Russian Public Discussion', *Environmental Politics*, 19: 179–195.

UNFCCC (2011) 'Joint Implementation', http://unfccc.int/kyoto_protocol/mechanisms/joint_implementation/items/1674.php, date accessed 7/05/2011.

Wilson Rowe, E. (2013) *Russian Climate Politics: When Science Meets Policy*, Basingstoke: Palgrave Macmillan.

World Bank (2008) 'Energoeffektivnost' v Rossii: Skrytyy Rezerv', www.ifc.org/ifcext/rsefp.nsf/AttachmentsByTitle/FINAL_EE_report_Engl.pdf/$FILE/Final_EE_report_engl.pdf, date accessed 13/07/2012.

Yablokov, A. (2010) 'The Environment and Politics in Russia', *Russian Analytical Digest*, 79: 2–4.

Yanitsky, O. (2009) 'The Shift of Environmental Debates in Russia', *Current Sociology*, 57/6: 747–766.

Yanitsky, O. (2011) 'Pozhary 2010 v Rossii: Ekosotsiologicheskiy Analiz', *Sotsiologicheskie Issledovaniya*, 3: 3–12.

Zakharov, V. (2011) 'Modernizatsiya Ekonomiki i Ustoychivoe Razvitie', *Na Puti k Ustoychivomu Razvitiyu Rossii*, 55: 11–15.

4   Russian newspapers and climate change

This chapter offers an analysis of Russian press coverage of climate change which allows us to identify the priority themes within the coverage, changes in these priorities and even omissions of certain events or facts. More importantly, it allows us to test for a correlation between state policy and media policy, one of the chief hypotheses in this project. Data were collected from five national newspapers: *Izvestiya, Kommersant, Rossiyskaya gazeta, Komsomol'skaya pravda* and *Sovets-kaya Rossiya*. The aim of the chapter is to study the dynamics of media coverage by looking at how the amount of climate change news depended on certain conditions over time (modifications in state policy, global conferences on climate change, acceptance of international documents and so on). Through discourse analysis, the chapter will also look at how the character of these articles varied under the specified conditions (how climate change and state policy on it were portrayed, who were the main newsmakers and opinion leaders on the topic and so on).

Coverage in the newspapers mentioned above will be studied by focusing on three events: the Kyoto Conference (1997), the Copenhagen Conference (plus acceptance of the Climate Doctrine) (2009) and the heatwave in Russia (2010). The rationale behind these choices is that state policy changed tremendously between the Kyoto and Copenhagen Conferences, so it allows us to see if there was a correlation between this change and media policy, whilst the heatwave allows us to explore whether other reasons, such as natural disasters, have more influence over the coverage than does state policy. The information collected will be used to provide an understanding of coverage. Then articles will be studied on the subject of how often they refer to official sources, NGOs, businesses and scientists in their information on climate change. Furthermore, samples from different newspapers and different events will be analysed by means of discourse analysis within specific politico-economic contexts. The hypotheses along with the choice of media organs and events will be explained before presenting and examining the results of the analysis.

Media analysis

Before the media analysis was performed, the following hypotheses were drawn:

1   Russian media organs owned by actors with an interest in continued carbon emissions will take a more sceptical/hostile view towards climate change and/or produce less coverage.

2   Russian media organs relying heavily on advertising by actors with an inter-
    est in continued carbon emissions will take a more sceptical/hostile view
    towards climate change and/or produce less coverage.
3   The Russian media will be 'drawn into a symbiotic relationship with power-
    ful sources of information by economic necessity and reciprocity of inter-
    est' (Herman and Chomsky 1994 [1988], p. 18). When either the sources of
    information change or those sources change their position, Russian media
    coverage will change accordingly.
4   Coverage of climate change will increase when it is in the interest of the
    'dominant elites' – for instance, after governmental acceptance of climate
    change regulation policy, ratification and approval of international docu-
    ments, participation in international negotiation and demonstration of a pro-
    active climate change mitigating policy.
5   Micro-factors of media production, such as journalists' professional norms or
    journalists' writing styles to approach the problem, do not result in major inter-
    ference and are subordinate to macro-factors such as ownership structure and
    dominant ideology. So, the coverage of climate change during extreme weather
    conditions (which arguably can satisfy journalists' desire to write about a sen-
    sational story) will not be greater than the coverage during the period of major
    international negotiations or national activity on climate change.

Testing of these hypotheses will allow us to see the influence of economic
and state elites (where they exist). Furthermore, if the chief hypothesis as stated
above (that of the dominant influence of the state over the Russian media) is correct, then in this case, coverage of climate change issues in Russia should change
depending on the state's policy. For instance, the coverage before the Copenhagen
Conference in December 2009 should not really be significant and qualitatively
it should be presented in a way which ridicules climate change, seeks to diminish
the belief that humans are to blame or maintains that Russia has nothing to do
with it. Just before the Copenhagen Conference, when the Russian government
changed its position towards being more sensitive to climate change and adopted
the Climate Doctrine, the quantity and quality of news should have changed. The
statements acknowledging human impact and the negative effect for the country
will become more frequent (further on in this chapter, more predictions about
the analysed events will be spelled out and tested). Nevertheless, the commercial
unpopularity of the topic among businesses (media owners or advertisers) will
still be an obstacle to the popularising of the problem, as will tight connections
between the Russian state and businesses (especially the energy sector).

For the purpose of testing all of these suggestions in this chapter, five Russian
newspapers will be analysed through content and discourse analysis. The timeframe
of the news coverage will be limited to two months around the selected events.

The choice of media organs

In this research project it is vital to consider characteristics of the analysed media
organs such as whether they have federal or regional distribution, the size of the

actual audience and what media organs are the most popular among the intellectual, business and political elites of the society, as well as who owns them, how much they depend on advertising revenue and how much they are financially independent.

One of the central challenges in gathering data for this research and conducting media analysis is to find adequate representatives of all Russian media. It should be noted that any selection will have some limitations. For instance, as mentioned above, it seems obvious to base the selection of studied material on its popularity and audience size. At the same time, some of the newspapers are read only by a small group of members of the intellectual or business elites, who cannot be ignored during the analysis.

In order to gain an insight into the popularity of the Russian media outlets and their actual audience size, the opinion polls of the Public Opinion Foundation (FOM) are helpful. The poll 'Mass Media: Preferable Channels of Information', conducted by FOM (2007), has demonstrated that Russian people are actively interested in news about Russian and world affairs. They get most of their information through national television (90 percent of respondents see TV as a major news source). National newspapers were in second place, with 30 percent of respondents citing them. Regional TV, national radio, local newspapers and local TV follow one another, with little difference of 1 or 2 percent (29, 26, 26 and 25 percent respectively). The Internet had quite a low position, with only 9 percent of respondents using it as their preferred channel of information. More recent data (FOM 2013) show that even though TV and newspapers remained in the same positions (with 89 percent and 27 percent of population preferring these media for news), the Internet's audience has expanded significantly, with up to 29 percent of Russians using it as a news source.[1] Considering these data, it would be desirable for this research project to analyse the representatives of the most popular categories such as national television and national newspapers as well as Internet news websites. However, because the popularity of the Internet in the years of the selected events (for the purpose of media analysis) is very different (being very unpopular in the late 1990s and becoming quite popular after 2009), Internet sources were omitted. Considering how much larger TV audiences tend to be, and the extent of television's influence over audiences, analysis of TV news could have benefited this study. Nevertheless, bearing in mind the time constraints of the project as well as the extremely limited amount of coverage of climate change on Russian television, this source of information is omitted, and only print media available through electronic databases are analysed.

Following the same logic of identifying the most popular media outlets amongst these two categories, let's go back to the poll's results. The top two popular newspapers are *Argumenty i fakty* [Arguments and Facts] and *Komsomol'skaya pravda* [Komsomol Truth],[2] both of which can be considered tabloids. However, as will be discussed later, the history of the *Komsomol'skaya pravda* makes it difficult to equate the newspaper to, let's say, British tabloids. The same goes for *Argumenty i fakty:* in the 1980s, it was one of the trendsetters in the *perestroika* movement. Even though *Argumenty i fakty*, according to this opinion poll, is one of the most popular newspapers, it will not be used in this research, since it is a weekly, whilst all other studied media are published on a daily basis.

The analysis of *Komsomol'skaya pravda*, due to its popularity, is important for understanding what kind of information about climate change the majority of Russia's population receive. However, smaller-circulation organs such as *Rossiyskaya gazeta, Izvestiya, Kommersant* [The Businessman] and *Sovetskaya Rossiya* (identified as the number one choice by 3 percent or fewer of respondents) cannot be ignored due to their special ownership structure, target audience or political affiliation. *Rossiyskaya gazeta* is an official newspaper of the Russian government, and its coverage heavily depends on the state's official policy. *Izvestiya* and *Kommersant* belong to the quality press and position themselves as independent, aimed at the so-called elite or decision makers: highly educated people, managers, politicians, members of the intelligentsia and so on. *Sovetskaya Rossiya*, in turn, is a left-wing[3] newspaper which is popular among senior citizens who proudly carry on the ideals and traditions of the Soviet legacy or use the newspaper as an arena to disagree with the current government policy or more generally with the modern capitalist world order.

So, the choice for the analysis of Russian print media organs are *Komsomol'skaya pravda, Rossiyskaya gazeta, Izvestiya, Kommersant* and *Sovetskaya Rossiya*. Relevant information about the chosen media outlets is presented below.

Komsomol'skaya pravda (KP)

Type: Newspaper – tabloid
Frequency: Daily
Circulation: 655,000; Friday issue: 3,000,000

Ownership structure: Belongs to the ESN group of companies (mostly concerned with energy production) (Media Atlas 2011) and has close ties with the Russian railway company, OAO RZhD, which is considered the second largest monopoly in Russia.

Additional information:
The majority of the readers are women. The newspaper is mostly read by married people in the age group 45 or older (Atlas SMI 2011). Even though now the main aim of the *KP* is entertainment, which includes capturing the audience's attention with coverage of various scandals and celebrity news, it was first established in 1925 as the main media organ for Soviet youth (Komsomol). By using the language of its direct audience (Strovskiy 2011) in less formal ways than other newspapers, it would spread news about the best representatives of the Soviet youth (true communists and hard workers who would build a better future for the country). Over time the *KP* became increasingly popular, and the newspaper not only acted as an outlet for the Party line, but also during the years of Khrushchev's thaw it heavily criticised the individual institutions of the Soviet government or people in charge of them. Strovskiy (2011) points out that at that period, the influence of the *KP* was really significant and even ministers and Party members were

afraid of its critique. After *perestroika*, the newspaper moved away from politics to the infotainment sphere. According to the BBC news website, 'it [*KP*] has built its reputation on a gentle nostalgia for the Soviet period, firm backing for Kremlin policy and a keen interest in celebrity news and scandal from home and abroad' (BBC News 2008), which might be explained by the newspaper's historical heritage and ownership structure dominated by the energy companies.

Rossiyskaya gazeta (RG)

Type: Official newspaper of the Russian government
Frequency: Daily
Circulation: 179,240
Ownership structure: Government of the Russian Federation

Additional information:
The majority of the readers are men. The newspaper is mostly read by married people with higher education in the age group 55 and older (Atlas SMI 2011). According to the official website of the newspaper, it 'enjoys official status, because acts of state come into effect upon their publication here' (Rossiyskaya gazeta website 2011). At the same time, the *RG* does not restrict itself to publishing only official documents, but also tries to attract the attention of the general reader by covering various types of domestic and international news. The newspaper defines its readership as an 'even-tempered adult inclined to conservative views'. Even though over time, the newspaper has published some criticism of some state institutions, it is expected to cover the state's official position in a manner which would appeal to the supporters of state policy.

Izvestiya

Type: Social-political and business newspaper
Frequency: Daily
Circulation: 234,500

Ownership structure: Until May 2008 the paper was owned by Gazprom (BBC News 2008). The latest information on the ownership of *Izvestiya* claims that it is part of the NMG media holding group, whose shares belong to OAO 'AB 'Rossiya' – 54.96 percent (co-owner Yuriy Kovalchyuk is widely reported to be a close associate of Vladimir Putin), OAO 'Surgutneft' 19.49 percent, OAO 'Severstal' 19.49 percent, the SOGAZ Insurance Group 6.06 percent (Media Atlas 2011).

Additional information:
The majority of the readers are men. The newspaper is mostly read by married people with higher education in the age group 65 and older (Atlas SMI 2011). *Izvestiya* is considered a centrist newspaper with a predominantly liberal readership. Like the *KP*, *Izvestiya* was also first founded at the birth of the Soviet era in 1917. At first just a mouthpiece of the Communist Party, it eventually became

popular among intellectuals and academics. During and after *perestroika*, it criticised the Kremlin's policies on various occasions; however, it has been noted that
since the newspaper had been bought by SOGAZ, its media policy went through
changes once again (BBC News 2008).

Kommersant

Type: National business newspaper
Frequency: Daily
Circulation: 125,000–130,000

Ownership structure: 1999–2006 belonged to oligarch Boris Berezovsky; in
2008 was bought by oligarch Alisher Usmanov (owner of Metalloinvest) (Online
Gazeta 2011).

Additional information:
Audience demographic characteristics: men 57 percent, women 43 percent,
managers 29 percent, professionals 19 percent, office workers 13 percent, workers 12 percent, students 4 percent (Kommersant website 2011). *Kommersant*
positions itself as a right-wing liberal newspaper, which is mostly aimed at businessmen or would-be businessmen. From the beginning, when it was founded in
1989, the newspaper was plotted as an analogue of the Western quality press, and
its articles would lack evaluations or judgements of the events described, sharing
only facts and information which would not interest ordinary readers but would
appeal to managers or specialists (Strovskiy 2011). It characterises itself as 'one
of the most authoritative and influential publications for Russia's decision makers' (Kommersant website 2011).

Sovetskaya Rossiya (SR)

Type: National left-wing (communist) newspaper
Frequency: 3 times a week
Circulation: 300,000

Ownership structure: Its editor, Valentin Chikin, on its official website in 2012,
claimed that 'economically the newspaper is independent of any power structures,
parties or financial groups. The basis of its budget comes from the subscription
and readers' donations. Profit from advertising is negligible'.

Additional information:
SR was launched in 1956 as the official press organ of the CPSU (RSFSR) and
the Council of Ministers of the RSFSR. In 1990–1991 its editor-in-chief turned
the newspaper into an opposition organ and channel to transmit the ideas of the
Russian Communist Party, which makes it (in this sample of media organs) a representative of the left-wing media. One of the specific features of this newspaper in
comparison with the other studied press organs is that three quarters of every issue
of *SR* consists of articles written by its readers, rather than professional journalists.[4]

According to the opinion poll taken by the Public Opinion Fund (FOM 1997) in Moscow, 61 percent of *SR* readers are retired people (by comparison, the same research for other newspapers showed on average 31 percent of the readership of all newspapers are seniors).[5] Also, in comparison with 'an average Muscovite', readers of *SR* are quite politicised: they show interest in political news, comments on political news and news from other regions twice as much as do other readerships.

Kommersant, KP, RG and *Izvestiya* were accessed through an online database: Public.ru, whilst *SR* was accessed through the database Integrum World Wide.[6] These organisations are commercial and collect mass media information in order to sell access to business, government and academic organisations so they can monitor their press ratings or conduct other research relevant to them. In order to find news/articles applicable for this project, a set of keywords was specified. They all referred to climate change in one sense or another. The options included: 'climate change', 'global warming/cooling', and the 'greenhouse effect' (изменение климата, глобальное потепление, глобальное похолодание, парниковый эффект). Depending on the analysed event, keywords included: the Kyoto Conference, the Kyoto Protocol, the Copenhagen Conference and the Climate Doctrine (Киотская конференция, Киотский протокол, Копенгагенская конференция, Климатическая доктрина).

The choice of events and expected results of media coverage

The media analysis will be of a two-month period covering three events related to climate change issues. The events selected are considered to be 'critical discourse moments'; as Carvalho (2008, p. 166) states, '[c]ritical discourse moments are periods that involve specific happenings, which may challenge the "established" discursive positions. Various factors may define these key moments: political activity, scientific findings or other socially relevant events'. In order to test the main hypothesis of Russian media dependency on Russian state climate policy, the choice of events is predominantly determined by what position towards climate change regulations at the specific point in history the Russian government held. The analysed events are:

1   The United Nations Climate Change Conference in Kyoto, Japan, 1–11 December 1997
2   The United Nations Climate Change Conference in Copenhagen, 6–18 December 2009 and acceptance of the Climate Doctrine of the Russian Federation on 17 December 2009
3   The heatwave in Russia in August 2010

The United Nations Climate Change Conference in Kyoto, Japan, 1–11 December 1997

The Third Conference of the Parties under the UNFCCC is famous for the announcement of the Kyoto Protocol, in which for the first time states agreed to legally restrict the amount of GHG emissions allowed for the signatories. The principle and significance of the Kyoto Protocol is discussed in detail in Chapter 3. For this section it is important to understand that in December 1997, Russia had a strong anti-Kyoto position and, together with the United States, Canada, New Zealand and Japan, was lobbying for lower emission restrictions for industrialised countries. Also at this time there was a strong negative attitude towards the Kyoto Protocol among the Russian scientific community and the president's advisory team (the opposition was led by economic adviser Andrey Illarionov and academic Yuriy Izrael). So, for this event the media coverage is expected to be low, since the climate change topic was very far from the state's agenda, and the character of the news is expected to be very negative. For instance, the origins of climate change should have been questioned: whether people have anything to do with it or not and whether it was happening at all. The Kyoto Protocol is expected to be seen as a danger to Russia's economic development, or even more as a conspiracy among the Western countries against Russia, with the view that Russia should abstain from signing it. The sources of information are expected to be dominated by government representatives and academics closely working with the Kremlin.

The United Nations Climate Change Conference in Copenhagen, 6–18 December 2009 and the Climate Doctrine, 17 December 2009

The Copenhagen Conference and Russia's position during its negotiation also have been discussed in detail in Chapter 3. The main issues connected with this event are that at this conference countries were supposed to come to an agreement and produce a new document to replace the Kyoto Protocol, which was supposed to expire after 2012. Eventually, for the global environmental and scientific community, which had big hopes for the conference, it turned out to be a fiasco, since major disagreements did not allow countries to finalise their decisions and produce a new legally binding document on climate change policy. In spite of the negative outcome of the conference, the Russian government announced its firm position that it would be working on reducing GHG emissions regardless of whether the global community came to an agreement or not. Closer to the end of the conference, the Russian Climate Doctrine was adopted, which officially stated that climate change had an anthropogenic character and that its negative consequences could damage the country's well-being.

Media coverage of climate change in this period of time is expected to mirror the official vision of the problem and accept the anthropogenic character of climate change and the position that mitigating climate change will benefit the country. In particular, the positive change in coverage should coincide with Medvedev's visit to Copenhagen and acceptance of the Climate Doctrine. The news mentioning the acceptance of the Climate Doctrine is expected to be slightly more extensive than that devoted only to the conference, since the doctrine was a direct initiative of the Russian government, whilst the conference until the very last moment was not favoured by the head of state.[7] However, due to the time overlap between these two events, it will be difficult to separate them. State officials are still expected to dominate among information sources. Representatives of NGOs and different academic institutes are also expected to act as sources, since the more government changes its policy towards climate change regulation, the closer it gets to the position of NGOs and scientists working on this problem.

Russian heatwave in August 2010

The anomalous heatwave of the summer of 2010 led to some very severe consequences in the central part of Russia. The temperatures in Moscow broke previous records. The heatwave provoked vast forest fires around major cities in central Russia, and official forces could not gain control for several weeks. As a consequence, extreme weather conditions moved climate-related topics up the news hierarchy. This event is interesting in two ways. Firstly, it happened after the president and government took the course of supporting climate change regulations after accepting its existence and its negative character for Russia. The media coverage, as in the previous case, should have been in favour of climate change regulation and mitigation. Secondly, if this natural disaster provoked much greater coverage than the state's participation in the UN conferences, then it would bring into question the argument of the dominant elites' influence.

If the chief hypothesis is correct, then the amount of articles during this period of time cannot be higher than during the Copenhagen Conference and it is expected to be consistent with the official position on the problem: that Russia is supporting climate change mitigation programmes and will take part in them. The information sources are expected to be the same as for the previous event, but with greater contributions from NGOs and academia (see the previous section for explanation).

These three events are good cases to consider in order to see how media coverage has changed along with the differing political contexts.

Content analysis of media coverage of climate change in Russia

The first stage of media analysis consists of content analysis, which considers the numbers of articles mentioning climate change in all the studied newspapers across three selected events. The results of the content analysis are presented in Table 4.1.

In order to test the first two hypotheses (whether the coverage depends on the media ownership structure or advertising policy), a series of two sample proportion tests was conducted (with Stata software). The test aimed to assess the statistical significance of differences among various newspapers within the studied timeframes. For instance, the proportion of articles published by *KP* during the Kyoto period was compared with the proportion of articles published by *Kommersant* within the same timeframe, then with the proportion of articles published by *RG* and so on (the test was run for all possible pairwise combinations). During the Kyoto period, the test did not show a significant difference over all five newspapers (p-value > 0.05), regardless of whether we are talking about the state-owned *RG* or the profit-oriented *KP*. During the Copenhagen Conference and heatwave

Table 4.1 Number of articles mentioning climate change within the cases studied

Newspaper	Kyoto Conference 1.11–31.12.1997	Copenhagen Conference 1.11–31.12.2009	Heatwave 1.07–31.08.2010
Komsomol'skaya pravda	**4*** (Moscow**) 0.2%***	**19** 0.54%	**13** 0.43%
Rossiyskaya gazeta	**2** (federal issue) 0.1%	**41** 0.84%	**32** 0.63%
Izvestiya	**3** (Moscow) 0.18%	**30** 0.55%	**22** 0.48%
Kommersant	**3** (Main) 0.1%	**22** 0.61%	**6** 0.18%
Sovetskaya Rossiya	**1** 0.2%	**15** 3%	**7** 1.4%
Total	**13**	**127**	**80**

* Absolute number of articles on climate change published within the studied timeframe in the selected newspapers.
** 'Moscow': articles were published in newspapers distributed in the Moscow area; for *Rossiyskaya gazeta* and *Izvestiya*: 'federal issue' includes articles distributed throughout the country; for '*Kommersant*': 'main' category excludes the articles published in specialised issues of the newspaper such as *Kommersant-den'gi* and so on.
*** Percentage of articles on climate change towards overall number of articles published within the studied timeframe in the selected newspapers.

periods, the communist newspaper *SR* significantly stood out, publishing more articles than the other newspapers (whether this significant difference in the number of published articles represents a different media policy on climate change by this media organ will be studied below through discourse analysis). Furthermore, during the summer of 2010, *Kommersant* published significantly fewer articles than the other newspapers.

The analysis demonstrated the following results: *Kommersant* and *RG* (p-value = 0.0034), *Kommersant* and *Izvestiya* (p-value = 0.0313).[8] As was mentioned earlier, *Kommersant* characterises itself as a quality newspaper for professionals, which tries to maintain its image as a serious business media outlet. As will be discussed below, the heatwave in 2010 produced mostly sensational articles, which might be a reason for *Kommersant*'s reserved coverage during this particular period (it is important that during other periods, it did not show this difference). The conducted analysis shows that the first two hypotheses were not quite borne out. Whilst the supposedly independent *SR*, indeed, published proportionately more articles on climate change, the other four newspapers did not show significant differences over all three periods (with only *Kommersant* being relatively reserved during the heatwave). The similarities as well as acknowledged differences will be analysed further through discourse analysis.

With regard to the fourth hypothesis, that media would follow the interests of the dominant elites, all newspapers did follow the same trend, where 12 years after the Kyoto Conference the coverage of climate change had changed tremendously, from almost no representation in 1997 (cumulative number of 13 articles) to some representation in 2009 during the Copenhagen Conference (127 articles).

The argument of the significant influence of the Russian government over media coverage can also be supported by the comparison of the last two events. Even though many argue that unusual weather conditions bring attention to climate change, the highly politicised events of December 2009 (Copenhagen Conference and the acceptance of the Climate Doctrine) provoked 47 articles more than the heatwave. However, it should be noted that in the Copenhagen case, there is no alteration of coverage before or after Medvedev's attendance at the conference or the acceptance of the Climate Doctrine. For instance, only three out of five newspapers mentioned the Climate Doctrine and only did this once (this finding is discussed in more detail later in this chapter).

The fifth hypothesis appears to be partially correct. One of the possible reasons for journalists' lack of interest in climate change is that it does not have direct relevance to the readership (for more see Chapter 1), so the events of summer 2010, with its direct effect on millions of Russians, could have been a 'perfect' situation for journalists, when climate change quite literally entered the houses of much of the readership. Indeed, the coverage of climate change at this time was relatively high (compared with the Kyoto coverage); however, it was still 37 percent less than during the politicised events of December 2009 and, as will be discussed in more detail below, in many cases it did not refer to the direct correlation between natural abnormalities and the climate change phenomenon. At the same time, coverage of the Copenhagen Conference and the Climate Doctrine were much clearer on climate change issues and their relevance to Russia's interests. This observation confirms that the decisive factor in coverage is whether the relevant elite groups (in Russia's case, the government) take an interest in an issue or not. This hypothesis will also be tested by means of discourse analysis, which will demonstrate to what extent coverage of the heatwave in 2010 was influenced by state policy. If the majority of articles still reflect state policy rather than trying to independently explore this issue, then the role of 'micro-factors' will be even more diminished in the case of Russian media.

The third hypothesis of the choice of information sources needs to be analysed in greater detail. The summary of the choice of sources appears below in Figures 4.1 and 4.2. The articles were categorised according to their types of sources, defined as: Russian official sources (ROS), foreign official sources (FOS), business, science, NGOs, international organisations (IO) and others.

During the Kyoto coverage, some references to foreign sources (businesses, scientists or international organisations) appeared, with few references to Russian official sources, which can be explained by the state's disinterest in climate change at that time. The coverage of the Copenhagen Conference and the heatwave was overall characterised by a high number of Russian official sources being used. Perhaps this is understandable with regard to the Copenhagen Conference and the Climate Doctrine, since these were highly politicised events (as was the Kyoto Conference; however, Russian officials remained quiet at that time). But even during the heatwave, a natural phenomenon, the state's officials were still in high demand as an information source, being second after 'Science'. Russian official sources were dominated by two specific figures – the president and prime minister. For example, in *Izvestiya*'s coverage of the Copenhagen Conference, Medvedev's words and position kept reappearing throughout the texts, where he

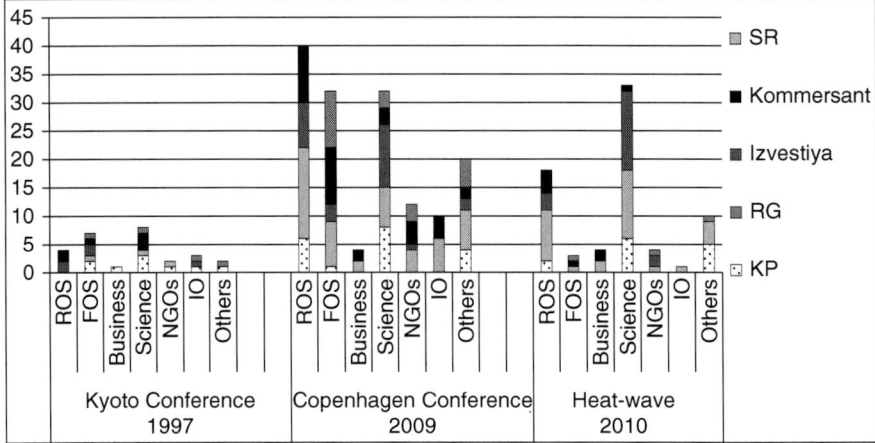

Figure 4.1 Number of 'sources of information' in the newspapers studied

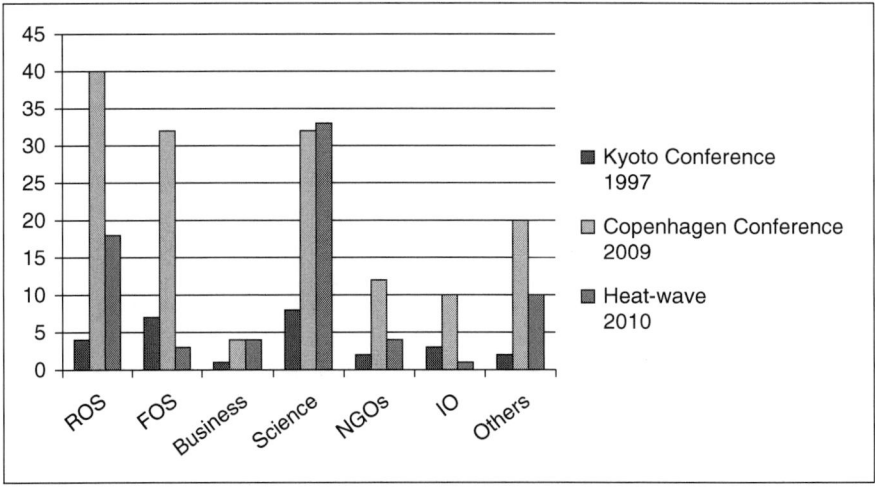

Figure 4.2 Total number of 'sources of information' over the three events

appeals to people not to panic, 'not to let them trick us' ('*ne dat' nas razvesti*') (Farizova 2009b) and not to forget about the country's national interests.

After the Russian government, the second most popular source across all events and all newspapers examined was 'Science', which included both Russian and foreign representatives. It should be noted that quite often it was either scientists of the hydrometeorological centre of Russia (the head of which for a long time was Alexander Bedritsky, the current president's adviser on climate change issues) or the general concept of scientists as such ('as scientists agreed', 'as many scientists think' and so on).

Another interesting observation which can be made is that overall there was extremely low reference to 'business' sources. This is strange, considering how much GHG emissions depend on the business sector and how much (in Russia's context) the gas and oil industries might suffer from climate change progression but also presumably from mitigation measures (see Chapter 3). There were also slight, but rather predictable, variations among the information sources of the different newspapers – for instance, *Kommersant* more often than others referred to 'business' sources, and *Komsomol'skaya pravda* more often used non-standard sources of information (for example, regular people as witnesses of climate change). In contrast to the other newspapers, *Sovetskaya Rossiya* made no references to Russian officials at all, but instead made frequent use of foreign state officials (such as Venezuela's president Hugo Chavez or Bolivian leader Evo Morales). Another peculiarity of *SR* was that quite often the authors of the articles were themselves significant figures with a very strong opinion on the subject (the leader of the Communist Party of the Russian Federation, Gennady Zyuganov, or the former leader of Cuba, Fidel Castro). In some ways they became both journalists and information sources in one.

Overall, the third prediction of the media's dependency on sources of information has proven to be correct. In particular, all newspapers (apart from *Sovetskaya Rossiya*) have demonstrated the correlation between the change in the state position on climate change and media dependence on the 'Russian officials' as sources of information. This finding was demonstrated not only by the number of mentions of Russian officials in the articles, but also the way they enter the discussion – as saviours of the negotiation process, the highest source of authority on the subject or as defendants of national interests.

Discourse analysis of media coverage of climate change in Russia

The methodology for the analysis of Russian articles on climate change is inspired by the approach suggested by Mautner (2008, p. 30) for the analysis of print media, which, as she mentions, 'draw[s] on the tradition of both critical discourse analysis and corpus linguistics'[9] and fits within the methodology proposed by van Dijk (1991, 2001, 2011). Even though the toolkit of Mautner's method suggests studying seven elements of the text, only six will be used for this research.[10] They are *lexis, transitivity, modality, sources, textual coherence* and *argumentative devices establishing rapport between author and reader* – each of these assesses the different levels of the discourse created by the texts.

Mautner states that 'on the level of lexis, the analyst will try to identify patterns in the choice of words, and in particular those with a distinctive "evaluative meaning"' (2008, p. 38). In the case of media coverage of climate change in Russia, it is also necessary to study what kind of words are used to describe climate change – whether it is referred to as a 'fact', 'lie, plot or fiction' and 'disaster or apocalypse' can create three very different pictures and messages for the audience. Carter and Simpson (1989, p. 290) state that 'transitivity [. . .] shows how speakers encode in their language their mental picture of reality and how they account for their experience of the world around them.' In order to analyse texts on the transitivity level, the following analysis looks at how climate change is pictured – for example, is it an uncontrollable natural force or a consequence of anthropogenic influence? How are decisions made to control climate change? Is Russia an active participant, an observer or a victim? Such analysis of subject–object relations will not only allow us to see who is perceived to be in charge of the situation, but also who is to be blamed for it, of course, if anyone (if climate change is indeed seen as an uncontrollable force, then it takes responsibility away from people).

Mautner 2008, citing Stubbs 1996 p. 202, defines *modality* as 'the ways in which language is used to encode meanings such as degrees of certainty and commitment, or alternatively vagueness and lack of commitment, personal beliefs versus generally accepted or taken for granted knowledge.' In the study of media coverage of climate change, this stage of analysis will be able to demonstrate whether the coverage is still sceptical and journalists are unsure about climate change's reality or its existence is no longer questioned. Textual coherence refers to the structure of a text, how certain information is emphasised through repetition or on the contrary omitted from the text, or two issues which are not obviously related are mentioned in the same text. In the case of climate change, the example of such textual linkage could be seen, for instance, in the articles where climate is mentioned along with other global problems, such as child labour, Somali pirates, education in the developing world and so on. In this case, even if climate change is not familiar to a reader, he can still make an educated guess that it is as important as other problems mentioned next to it.

The last level is the level of argumentative devices, which show the relations between the author and his audience. Mautner (2008, p. 43) states that these relations can be achieved 'through the use of rhetorical questions [. . .] appealing to the supposedly unifying force of common sense, [. . .] and the construction of a "we" group.' Indeed, in the Russian case the journalist's appeal to the common sense of the audience is quite frequent (for example, 'it is obvious that any kind of warming in Russia's severe climate will only benefit the country'), which implies that the author and the audience are in full agreement with each other and there is no room for challenging the journalist's statement. Another common demonstration of the unification of the author's and the reader's attitudes towards the climate change problem in Russia is the repetitive use of the pronoun 'we' ('we all know', 'we saw').

Further on, the analysis will be applied to the collected data. It should be mentioned that only articles which discuss climate change as one of the central topics will be analysed in this section. Approximately half of the studied articles mention the keywords used in a context not quite related to climate change, and the overall content of the articles was devoted to something either unrelated (for example, a celebrity mentioning climate change as one of her fears) or related in a distant way (for example, government negotiations mentioning the Kyoto Protocol as one of several forms of cooperation). Hence, for discourse analysis, articles which do not just mention but also discuss the problems of climate change in some detail were selected. Table 4.2 presents a summary of the main themes underlined by the discourse analysis of the data collected, followed by a detailed case-by-case analysis.[11]

Table 4.2 Main themes in the media coverage devoted to climate change (CC)

	Kyoto Conference	Copenhagen Conference	Heatwave
KP	Conference = 'battle'. Alarmist messages. Surprise of Russia's successful trend of reducing GHG emissions. Sarcastic way of portraying Russia's successes in GHG reductions. No mentioning of Russia's resistance to the Kyoto Protocol.	Variety of topics. Choice of words with high emotional value. Journalists – 'one of the people'. Russia – a leader of the climate change negotiation process. Recurring theme of a 'Western plot' against Russia's interests. A vague picture of CC's nature and consequences. 'Win-win' situation.	Sensational nature of the heatwave and its consequences. Questioned the link between CC and the heatwave. Impossible to stop CC – the economy 'must' develop. Greater attention to the idea of a 'climate weapon'.
RG	N/A	Russia – a leader in the negotiating process. Russia's stable position vs. the chaotic behaviour of other states. Confirmation of CC existence. Sometimes its anthropogenic character is questioned. CC pictured from an upbeat position. Benefits for Russia.	Direct links between CC and the heatwave. Confirmation of CC. Some uncertainty of its consequences or nature. No conspiracy theories. Issue of public opinion. Government was prepared. The Kyoto Protocol – benefits and possibilities for Russia.
Izvestiya	Conference = 'battle'. Alarmist messages. Surprise of Russia's successful trend (so far) of reducing GHG emissions. Russian government – defenders of national interests. No mention of Russia's resistance to the Kyoto Protocol.	Questioned the existence of CC or its anthropogenic character. CC – 'a plot?'. Russia's firm position on CC and its leading role in the negotiations. Other countries cause difficulties on the way of reaching an agreement.	Almost none of the articles discuss CC to a great extent. 'Horrifying' consequences that the weather brought upon Russians. CC – one of the possible explanations of the heatwave. Existence of CC was not certain. CC's anthropogenic origin was brought into question even more. More sensationalist messages. Questions the Kyoto Protocol.

(Continued)

Table 4.2 (Continued)

	Kyoto Conference	Copenhagen Conference	Heatwave
Kommersant	Conference = 'battle'. Alarming messages. Scepticism towards anthropogenic CC. Russia 'the most sensible' position. US hostile position.	Economic aspect of the problem. 'Win-win' situation. Impersonal style (reporting of facts). A number of articles shared the climate sceptic position. Questions the activities of the Russian government (but not its leaders!).	Economic context. Resemblance with the business report. Criticisms of the government. President's behaviour – 'absolutely correct'. 'Win-win' situation.
SR	No reference to Russia. 'Main polluter of the planet' – the US (criticism of its citizens and government).	Destructible nature of capitalism – cause of CC. Heroes (e.g., Latin America) vs. villains (e.g., the US). Fidel Castro – author of several articles. No references to Russia or the Russian government.	None of the articles discussed climate as a central topic. Leader of the Communist Party blames the party in power for burning Russia. Main reason for social-environmental apocalypse – the current financial system. CC – a definite threat, a danger for the whole world and everyone should unite to fight it. No direct mention of specific Russian state officials.

The Kyoto Conference coverage

The limited amount of articles devoted to climate change during this period of time does not really present much material for analysis. The omission of the topic speaks louder about the attitude towards climate change problems. However, applying the methodology described above to the collected data, some conclusions can be drawn. For *Kommersant, Izvestiya, KP* and *SR* (*RG* only mentioned climate change among other scientific topics discussed in its articles), the Kyoto Conference became the starting point for the articles. All newspapers compared the conference with some kind of a battle or a scandal – 'scientists acted on the offensive' (Kabannikov and Potapov 1997), 'general abuse', 'arguments continue' (Golovnin 1997), 'heavyweights' (*Izvestiya* 1997) or 'fierce disputes' (Motskobili 1997). *Izvestiya, KP* and *Kommersant* reproduced the alarmist messages which aimed to raise major concerns among their readers – 'God, what do we inhale!' (Chizhikov 1997), 'Japan will sink' (Kabannikov and Potapov 1997) or 'we are all hostages in climate thriller' (Motskobili and Maksimenko 1997).

In both *Izvestiya* and *KP* there was the reappearing theme of the surprise at Russia's (so far successful) trend of reducing GHG emissions. Whilst the *KP* talks about Russia in a more sarcastic way by pointing out that reduction has happened unintentionally due to the economic decline of the 1990s, *Izvestiya* pictures the Russian government as defenders of national interests which is trying to achieve mutually beneficial results (*Izvestiya* 1997). *Kommersant* also praised Russia's 'realistic' goals (Motskobili and Maksimenko 1997) and very 'sensible approach' to the problem (Motskobili 1997). Interestingly, whilst one of *Kommersant*'s articles provides a very clear and rational description of the climate change problem and its anthropogenic character with even slightly alarming messages of climate change's negative consequences (Motskobili and Maksimenko 1997), another article purposefully debunks the anthropogenic character of climate change (Shvarts 1997). Even the title states that 'there is no one to blame' (*'kraynikh net'*), the article continues with arguments against industries' impact on climate change, supporting this statement by the opinion of scientists at Harvard University.

In the *SR*'s article, on the contrary, there was no reference to Russia; instead, the article was devoted to the 'main polluter of the planet' (the United States), whose citizens lack 'environmental awareness' and whose government refuses to follow through on its initiatives to develop sustainable energy sources (Popov 1997).

Conclusion on the Kyoto Conference coverage

To sum up, the coverage of climate change during the Kyoto Protocol was low, as expected, so in this sense the earlier prediction proved to be correct. Few articles published at the time in the studied newspapers gave an adequate picture of the problem, rather they speculated about the sensational features of it (disastrous nature of climate change or the political fight around climate change). Russia's performance was never questioned and it was not articulated how unwilling Russian authorities were in their cooperation on mitigating climate change. On the contrary Russia was portrayed as a leader of the negotiating process, or in the case of the oppositional media organ *SR*, it was absolutely ignored and instead attention was diverted towards the United States. It could be argued again that the mass media were defending the state's position or at least not challenging it. At the same time, rather than coverage of climate change being biased or sceptical, the real issue was a striking overall absence of information on climate change.

The Copenhagen Conference and Climate Doctrine coverage[12]

Izvestiya

The reason for *Izvestiya* publishing the articles during this period of time became either the conference (predominantly) or the initiatives by Medvedev or the Russian government. All articles could be roughly divided into two groups. The first one questioned and debated the existence of climate change or its anthropogenic character. In this case, the article 'Osnovy naturfilosofii' in *Izvestiya* (Sokolov 2009) was especially interesting, given its disbelief in the existence of climate change backed up by referring to it as a matter of faith and the negotiations on climate change as religious wars. Also, quite often in these articles journalists appeal to the abstract universal concept of 'common sense' or confirm their position by using the term 'opinion of many scientists'. The second group is devoted to the report of Russia's firm position on climate change and its leading role in the negotiations (Farizova 2009b), whilst other countries were described as the troublemakers which slowed down acceptance of the agreement.

Komsomol'skaya pravda

In contrast to *Izvestiya*, the *KP* had a greater variety of reasons for the articles to be written, including the conference, the government's initiatives, weather anomalies, opinions of climate sceptics and even competition for readers which the newspaper initiated. Probably due to its tabloid nature, the style was also quite different from other newspapers, and the choice of words often had higher emotional value: 'disaster film "Day After Tomorrow" might become a reality' (Moiseenko 2009), 'major myth of 21st century' (Kovyneva and Moiseenko 2009), 'unbelievable natural abnormality' (Smirnova 2009) and so on. Furthermore, journalists tried to represent themselves as 'one of the people', which could be observed through the frequent repetition of the personal pronoun 'we', and to engage their readers by appealing to them through questions, which were frequently used as titles of the articles – 'Have Russian Hackers Exposed the Myth of Global Warming?' (Kovyneva and Moiseenko 2009) or 'Global Warming – Climatologists' Fiction?' (Smirnova 2009) and so on.

Like *Izvestiya, KP* also depicted Russia as a leader of the climate change negotiation process, whilst Medvedev was described as one of the saviours who would prevent the 'climate catastrophe'. Despite all disagreements in the scientific and international communities, the Russian president reassured his citizens that there was no reason to panic and the country would benefit from the situation in any case even if 'God forbid, [. . .] climate change is really happening' (Krivyakina 2009). There was also the recurring theme of a 'Western plot', where Western countries were trying to exploit the rest of the world and Western scientists were exaggerating the seriousness of the problem. Overall, the discourse created by the articles of *KP* drew a vague picture of climate change, where it was not really confirmed whether it was happening or not and to what extent, and where only the Russian government's position was clear, and in any case would not damage the country's national interests.

Kommersant

The discourse created by *Kommersant* slightly differed from the other newspapers studied, mainly because in most articles, climate change issues were discussed from the economic perspective where GHG emissions became another currency in the modern world: 'global warming was announced as profitable for Russia' (Granik 2009b). Another peculiarity of this newspaper was its generally impersonal style, where the main role of journalists was supposedly to present the facts and describe events from different points of view.

There were also two articles which shared the climate sceptic position and even referred to the opinion of Illarionov, whose role in Russia's anti–climate change position has been discussed earlier ('British Scientists Underestimated Russian Climate' [Sapozhkov and Butrin 2009] – the article was provoked by the Climategate scandal), or made fun of this environmental abnormality ('[e]ven if climate change does not exist, we still had to invent it, so hundreds of bureaucrats were able to spend the state budget on their crusade against cars' emissions' [Kharnas 2009]). Another interesting feature of this newspaper was that, like others, *Kommersant* quite often referred to the state in discussing climate change problems. But more than other newspapers, it tried to question the activities of the Russian government, perhaps not so much its chief executives (which also was present here but in a more indirect way – for example, through the sarcastic choice of the words '*Medvedev pogody ne sdelal*', meaning 'Medvedev did not play a role'[13]) (Granik 2009a), but the civil servants and ministries were criticised quite openly (for example, in the article 'Protocol of the Missed Opportunities' (Shapovalov 2009), and the bureaucracy was blamed for causing insurmountable difficulties for business to be able to use benefits from selling unused GHG quotas.

Rossiyskaya gazeta

Predictably, *Rossiyskaya gazeta* in its coverage of climate change during the period studied was following the steps of the Russian government. Even more than in the previous newspapers, Russia was presented as a world leader in the negotiating process ('in GHG reductions, Russia is already a world leader' [Petrov 2009], 'Russia is extremely interested in concluding a new agreement' [Merinov 2009]). Another concept which is promoted by these articles is Russia's stable position against the background of the chaos of the Copenhagen Conference which was created by other participants. In several articles, after journalists described the battle between the rich North and poor South, or the emotional behaviour of the Latin American leaders, or the hopelessness of the conference organisers, there followed a paragraph which stated the clear position of the Russian president – 'we are committed to the GHG reduction process, but we will protect national interests' (Elkov 2009). Another distinguishing feature of the *RG*'s texts was that even though they almost always confirmed the existence of climate change, in some cases its anthropogenic character was called into question.

On average it pictured climate change from an upbeat position ('solutions are possible', 'existing technologies are enough') and as a process full of positive opportunities for Russia ('potential', 'benefits', 'opportunities', 'investments') (see e.g., Shmeleva 2009). Again, as in other newspapers, the idea that Russia would benefit in any scenario kept recurring. One of the most interesting and unusual articles was written by the former president of the USSR Mikhail Gorbachev (2009). This piece ('Kto Povyshaet Gradus?' [Who Is Raising the Temperature?]) was full of very emotional appeals to solve the problem and that it should be done by the states that were the most responsible for it. He blamed the current economic system ('irresponsible race for extra profit at any price', 'blind faith in the invisible hand of the market' and 'states' inaction') for the catastrophic environmental situation. He saw the solution in an intellectual breakthrough, moral re-education of business and the active role of civil society, but what was more important was that state leaders should become 'real' leaders.

Sovetskaya Rossiya

As in the previous case, the character of this newspaper's coverage could be predicted, of course in its own way, different from *Rossiyskaya gazeta*'s. The common theme of all the articles was the destructible nature of capitalism, which had brought climate catastrophes upon us. In the invective writing style, the authors tried to accentuate who was a hero (normally a representative of Latin America) and who was a villain (normally American or western European politicians). In one out of several articles written by Fidel Castro (2009a), already in its headline 'The Truth about What Happened at the Summit' ['*Pravda o Tom, Chto Proizoshlo na Sammite*'] implied that everything people heard before was not quite correct. The article was written in a very emotional manner, with frequent use of such strong phrases as 'before we discuss in what type of society we will live, now we discuss whether humanity will live at all', 'the last thing that people can lose is hope', 'men and women armed with truth' and so on. The speech of Barack Obama was described as 'deceptive and demagogic', whilst Evo Morales and Hugo Chavez both produced 'wise and meaningful' speeches which would be remembered in history as 'concise and relevant'. Then Latin American leaders completed their mission at the conference and Barack Obama, on the contrary, left before its completion. These remarks followed with the description of the 'amazing battle' of countries of the third world to rebell against the attempt of Obama and other 'rich countries' representatives' to impose the document proposed by the United States.

The peculiarity of *Sovetskaya Rossiya*'s authorship was discussed earlier on in the section on the information sources; however, this factor had a significant influence over the coverage of climate change in this newspaper. Fidel Castro was an author of three out of nine articles that discussed climate change as a central topic, and seven out of all fifteen articles published in the *SR* over the studied period with some mention of climate change. There is no publicly available information on how exactly the *SR* got hold of his articles, whether he writes specifically for them or they just reprint them from somewhere else. What is more important is that in his articles he is not just an ordinary journalist who passes information to his readers, but he shares his very active political position, he appeals to his fellow comrades and tries to disgrace his ideological opponents. In another of Castro's articles, 'Chas Istiny' [Moment of Truth] (Castro 2009b), he concluded with a highly emotional statement: 'For the heads of empires, in spite of their cynical lie, the time of truth comes. Their own allies trust them less and less. In Mexico [at the time of the publication, it was the next location of the UN climate change conference] as in Copenhagen and as in any other country, they will encounter growing peoples' resistance from those who have not lost their hope to survive.' Other articles were slightly more neutral but still stressed the opposition between the developed and developing countries and paid great attention to regular protesters who were manhandled by the Danish police.

Another very interesting point, especially in contrast to the coverage of all the other studied newspapers, was that in the *SR* there were again no references to Russia or the Russian government at all, in either a positive or a negative way, which could be seen as extremely odd. First of all, as an oppositional newspaper, *Sovetskaya Rossiya* could have used this opportunity to point out the destructive policy of current Russian officials or, on the other hand, to praise Russia's natural resources, such as boreal forest, and appeal to it as a possible environmental leader (but perhaps under a different government). One might argue that it is the former policy of the Soviet Union that brought Russia to be one of the greatest CO_2 emitters in the world and that the Soviet legacy of environmental neglect further stimulates Russia's environmental degradation.

Conclusion on media coverage of the Copenhagen Conference and the Climate Doctrine

The picture created by the study of the five newspapers is that they differ in their natures, such as *Kommersant* more often portraying climate change from the economic point of view; the *KP* bringing out the sensational nature of the problem; *Izvestiya* sharing a slight scepticism about climate change but mostly backing up the government's position; the *RG*, following the tradition of the Communist newspapers, praising the state's leaders; and the actual communist newspaper which currently is in opposition – *Sovetskaya Rossiya* – using this opportunity to underline the destructive nature of capitalism.

There were overarching themes throughout all the articles and no major contradictions between them were observed. All newspapers contributed to the creation of the same discourse, or it could be argued that they were all influenced by the same political, economic and social discourses. The picture supports the earlier prediction about the media coverage of these events. All newspapers to some extent did mirror state policy on climate change, and in the majority of the articles, climate change's existence was accepted, as was its anthropogenic origin.

One of the earlier stated hypotheses did not prove to be correct: the coverage after the acceptance of the Climate Doctrine did not change significantly, though three out of five newspapers (*KP, Izvestiya* and *RG*) mentioned its acceptance and its positive influence. It could be explained by the fact that the change in the state's policy on climate change did not happen during the Copenhagen Conference as such, but some time before it (after Medvedev's appointment to office); hence, the Climate Doctrine or Medvedev's speech at Copenhagen did not signify the start of the policy but was a logical step in its continuation.

Overall, there were some disagreements in the coverage and not all articles fully concurred with the above-mentioned statements; however, on average, similarities prevailed. None of the articles openly criticised the state's policy on climate change. Even when *Kommersant* did try to point out some drawbacks, the victim of its critique became the country's bureaucratic apparatus instead of its leaders. *Sovetskaya Rossiya*, which actively criticised the US and western European climate policies, did not mention the Russian government at all and did not use this opportunity to demonstrate its oppositional nature. Furthermore, in the majority of the articles (except for those in the *SR*), Russia was portrayed as a leader in the negotiating process, and it was noted how much the national economy could benefit from it. In contrast, there was hardly any mention that it was not long ago that the 'environmental leader' was one of the major obstacles to ratification of the Kyoto Protocol, and signed it due only to the certain benefits it was promised.

The heatwave coverage

Izvestiya

With only five articles directly related to climate change issues, almost none of them were actually about climate change. Most of the time, discussion started with the 'horrifying' consequences that the weather brought upon Russians ('Temperature records were broken' [Obraztsov 2010a], 'Why do we hear more often messages from the Ministry of Emergency Situations?' [*Izvestiya* 2010], 'climate of mass destruction' [Obraztsov 2010b]), whilst climate change would enter the articles further on in the texts as just one of the possible reasons which were constantly emphasised ('surely, global warming is only one of the reasons' [*Izvestiya* 2010], 'it would be too banal to explain temperature rise by global warming' [Obraztsov 2010a]).

Climate change's anthropogenic origin was brought into question even more ('climatologists still do not offer one explanation about climate change' [*Izvestiya* 2010], 'it is unclear who should be blamed for climate change – Earth or human beings' [Savinykh 2010]). Journalists tried to explore different theories about why the heatwave happened: for example, an increase in solar activity or even such an extraordinary one as a 'climate weapon', which explored the old concept of the Cold War where the United States was trying to destroy Russia's well-being. Not only were the reasons of the heatwave disputed but also the ways to cope with it; for instance, there were some questionable proposals for 'geoengineering' (Obraztsov 2010b). Russia's decision to join the Kyoto Protocol was discussed in the context that it did not require anything from the country and even promised some benefits; however, *Izvestiya* stated that how 'right' the decision was remained to be seen (Obraztsov 2010b).

Kommersant

For *Kommersant*, only two articles were distinguished which did not just mention climate change but discussed it in a more significant way. One of the articles talked about climate change in the economic context, mentioning the difficulties associated with the approvals of the Joint Implementation projects in Russia (Shapovalov 2010). The article resembled a business report, full of economic terminology and analysis. It should be noticed that the article did offer some criticisms of the government. For instance, it talked about the 'negative expectations' businesses had that projects that were approved would go to the major state corporations; however, the author noticed that these 'negative expectations were met only partially'. Whilst some approved projects were presented by such major companies as Gazprom Oil and Rosneft, they were 'diluted' by smaller projects.

A second article, titled 'Summer Will Call Us to Account' [*Leto Sprosit Strogo*] (Sborov 2010), connected the heatwave and climate change but also mentioned it in the economic context. The author quotes the presidential adviser on climate change, Alexander Bedritsky: 'if climate risks are miscalculated, then economic losses are unavoidable'. The article goes on to discuss whether the Russian Ministry of Energy should budget the possible risks of natural disasters (such as the heatwave of 2010) each year. It concludes with information on insurance companies which do not take into consideration the problem of global warming and will not adjust their tariffs.

Rossiyskaya gazeta

Overall, although the texts were full of very strong statements confirming climate change and its anthropogenic character (the 'majority of scientists confirm that global warming is happening [. . .] anthropogenic influence objectively contributes to climate change' [Elkov 2010]) and discussions about the consequences for Russia ('Russia is huge, so climate change will show itself in different ways' [*Rossiyskaya gazeta* 2010]), there was some place for uncertainty. In this case, it is interesting to look at the interview with Izrael (a prominent Russian clima-tologist and infamous climate sceptic). Izrael denied the connection between the heatwave and climate change but admitted that climate change was happening; however, there was no reason to worry – 'The process of melting will not take decades like the authors of environmental horror stories claim, but thousands of years. In this time, I think, human beings will find a solution' (Medvedev 2010).

One of the leading themes in this coverage was a demonstration of the possibil-ity that Russia could take a leading role in finding solutions to the problem ('Rus-sia should lead by example' [Shmeleva 2010]), and climate change was often accompanied by a discussion of national interests ('modernisation and energy efficiency of the Russian economy' [Pertsovskiy 2010]).

The *RG* raised the issue of public opinion by stating that the heatwave had managed to change the low level of awareness of the problem; however, the gov-ernment was prepared for it (interview with Minister of Natural Resources and Environment Trutnev: 'I reported to the government about possible threats[. . . .] I myself of course believed scientists, but there was a hope maybe it would not hap-pen tomorrow' [Smol'yakova 2010]). The Kyoto Protocol was mostly discussed in the context of the benefits and possibilities for the country ('a tool for moderni-sation and energy efficiency of the Russian economy', 'number of opportunities' [see e.g., Pertsovskiy 2010]).

Komsomol'skaya pravda

Due to the nature of the *KP* (being a popular tabloid), as in other cases, it often uses highly emotive words and sensationalist expressions ('Why did nature decide to fry us?', 'We are getting fried by a gigantic anti-cyclone – atmosphere mon-ster' [Smirnova 2010]), while some articles also pointed out the sensational conse-quences of climate change ('jellyfish in the Moscow River' [Mironov 2010]). Like other newspapers, the *KP* questioned the direct link between climate change and the heatwave (*Komsomol'skaya pravda* 2010), then in the same piece it would jump to the conclusion that even though people do influence the climate, it is impossible to stop climate change, since the economy 'must' develop. The coverage of the *KP* also paid greater attention than other newspapers to the idea of a climate weapon, devoting a whole article to looking at the different aspects of its possibility and only at the end in a brief paragraph mentioning a counter-argument (Kuzina 2010).

One out of the six articles devoted to climate change covered Vladimir Putin's trip to the far north and his meeting with scientists studying permafrost. The arti-cle showed Putin's concern with climate change problems: 'I just saw how fast the sea is "eating" the land, this is really impressive. However, nobody explained to me whether these changes are connected with human beings' influence'. After scientists reassured the prime minister (at that time) of the anthropogenic charac-ter of climate change, Putin shared one of his 'typical' climate jokes:

A thousand years ago mammoths started to die out around these territories, it is said, that it was connected with global warming and the shrinkage of the food supply. So mammoths aggregated on these islands. It all was without any anthropogenic character! It would be good if you [scientists] would tell us what is going to happen not only here in Russia, but also around the world, to which islands we need to migrate.

(Gamov 2010)

The author of the article remarks after this quotation of Putin that 'everyone understood it was a joke and had a laugh. Putin also laughed with everyone: "[W]hen will we be flooded or get frozen? Tell us in advance, so we know where to run", continued to joke Putin' (Gamov 2010).

Sovetskaya Rossiya

SR published seven articles during the studied period mentioning climate change, and even though none of them discussed climate change as a central topic, there are still some valuable observations to be made. Gennady Zyuganov (2010), leader of the Communist Party of the Russian Federation, wrote an article 'Politics of Catastrophes' ['*Politika Katastrof*'], where he argues that the climate, peat development or regular people were not the ones to be blamed for fires, but the current ruling party, whose 'incompetence' was burning Russia. Another article went even further and firmly stated that the main reason for social-environmental apocalypse was the current financial system, which must be destroyed. Climate change was described as a definite threat and a danger for the whole world, and everyone should unite to fight it (Khanzhin and Khromov 2010).

There were also three articles written by Fidel Castro where once again he mentioned that problems such as climate change were the consequences of neo-liberalism (Castro 2010). Although in *SR* coverage, there were some aggressive statements about state leaders, the current financial system and even the Russian government (Kramich 2010), including reference to 'soulless' bureaucrats and businessmen who were responsible for climate change due to their interest in immediate profit, once again there was no direct mention of specific Russian state officials or criticism of them.

Conclusion on mass media coverage of the heatwave

Once again the media coverage was influenced by the newspapers' defining characteristics. *Kommersant* looked at the economic aspect of the problem. The *KP* highlighted in its coverage the sensational facts. *Izvestiya* stayed quite sceptical towards the problem and at the same time sensationalist. The *RG* pursued a clear line of relations between climate change and the heatwave and used it as evidence of the right decision made by the government admitting the existence of climate change. The *SR* did not have any articles on climate change as a central topic, and in the articles where climate change was at least mentioned, the *SR* again blamed the current world political and financial system for the global problem.

Again common conclusions can be drawn. In support of the predictions stated above, overall there was much less uncertainty on reporting about climate change, since it became more apparent and more vivid, to the extent that for the first time *RG* mentioned Russian public opinion on climate change and in particular how low the level of concern was, and suggested that such unpleasant abnormal acts of nature might help to change this situation. All newspapers apart from *SR* pronounced the state's position and its decisions correct on the grounds that they would benefit Russia whatever happened. Once more, only *Kommersant* openly questioned the government's performance on climate change problems; however, it was not concerned with the country's leaders but ministers or other officials, and the *SR* questioned the capitalist policy in general. It should be noted that not all of the articles on climate change during this time were provoked by the heat-wave, but on the contrary some dealt with unrelated issues such as the implementation of the Kyoto Protocol.

Another interesting observation was made that was not apparent and noted in the predictions. It seems that the media coverage of climate change during the heatwave was much less structured and adequate[14] than it was during the time of the Copenhagen Conference and the Climate Doctrine. The issues which prevailed before in the political and economic contexts were raised again, such as uncertainty about the anthropogenic character and climate change's positive consequences for Russia. Furthermore, it provided more opportunities for sensationalist and alarmist messages, such as the climate weapon explanation of the heatwave (a conspiracy against Russia).

Discussion

The Kyoto and Copenhagen Conferences happened twelve years apart, and in this period of time Russia's state policy went through drastic changes. For a number of years it was sceptical of climate change and tried to stay out of international negotiations; at the same time, Russia had a strong anti–climate change lobby. A decade later, Russia changed its priorities and, just at the time of the Copenhagen Conference, accepted at the state level climate change's anthropogenic character and admitted that it was in the country's interests to take action on it. A limited selection of media organs was taken into consideration; due to practical reasons, this was an optimal way of conducting the analysis. For future research it could be interesting to study not only print media, but also TV, radio and Internet sources, and perhaps rather than selecting certain events, looking at a longer duration might reveal other interesting conclusions.

The analysis conducted showed that in the Russian case, the media coverage of climate change has indeed stayed within certain boundaries. One of the most striking findings demonstrated that the coverage of climate change in the studied newspapers has changed throughout the years from 13 articles during the Kyoto Conference to 127 at the time of the Copenhagen Conference. Even though during the heatwave the coverage stayed pretty high, with 80 articles, the event attracted less attention than did political affairs. It can be concluded that in the Russian case of climate change coverage, the state acts as an independent variable by directly or indirectly altering the media policy on the subject. Neither different ownership structures nor the degree of advertising dependency of Russian newspapers had much influence over the newspaper coverage. Even during the scarce coverage of the Kyoto Conference and especially during the Copenhagen Conference, the Russian state was presented in a beneficial way as a saviour of the negotiating process and Russian leaders were praised for their tactical approach to benefit the Russian economy in any scenario. Overall, the difference in the amount of articles between the newspapers was insignificant, with a few exceptions.

Kommersant published fewer articles on climate change during the heatwave. Perhaps it could be argued that the lower coverage in this newspaper can be explained by its ownership structure – Alisher Usmanov (Metalloinvest) owns it. Hence, it could be said that due to the industrial interests of the owner, *Kommersant* limited the coverage of climate change; however, this does not explain why *Kommersant*'s coverage did not differ in its amount during the other two events. Hence, we can argue that the explanation is in the newspaper's writing style, which tries to present information in a rational, more businesslike manner (which is also supported through discourse analysis of its articles). Objectively speaking, even though scientists do say that climate change makes extreme weather more likely, they are generally reluctant to assert a causal link in particular cases.[15]

The left-wing *Sovetskaya Rossiya* published more articles than the other newspapers during the time of the Copenhagen Conference and heatwave (if we look at how many articles it produced in proportion to the overall number of its news items per newspaper). Interestingly enough, despite the quantitative difference, *qualitatively* it also confirmed the Russian media's conformist position towards the Russian government with regard to climate change topics. Being a left-wing newspaper with strong support from Russia's Communist Party (the opposition party), it in fact presented a vast critique of the capitalist world and in particular blamed US president Barack Obama for all the problems. However, the *SR* did not exploit this opportunity to condemn Russia's government for its quite questionable climate policy or, on the other hand, for its rapid shift. So, indeed, the *SR* coverage differs from other newspapers, but it still stays within the 'manufactured consent' produced by the surrounding political, economic and social discourses.

These results could be explained, firstly, by the fact that the environmental degradation which Russia is currently facing is to a large extent a legacy of the Soviet Union's policies (see Chapter 3), and, secondly, by a recognition that there is a problem of political opposition and its role in Russia. For instance, Luke March (2009) explains how the Russian government 'manages the opposition' by 'creating' political parties (he provides an example of 'Just Russia'). Richard Sakwa (2011, p. 526) states that in Russia:

> [T]he role of political opposition is marginalized. Parties have limited political reach and fail to provide the framework for the institutionalization of political competition or the integration of regional and national politics. They are not the source of governmental formation, personnel appointments or policy generation; neither are they, more broadly, 'system-forming', in the sense of providing the framework for political order.

With regard to the Communist Party of the Russian Federation, Sakwa (2008) argues that it 'deserves the title of "party"' the most. However, he spells out a number of key problems within the Communist Party itself. For example, for a number of years, it has struggled with its own political identity, political goals and position towards the state leaders and ruling elites ('Do they need to overthrow them?', 'How compatible is party policy with the free market economy and capitalism?' and so on).

The analysis shows that after the change in the state's policy, in the majority of cases newspapers relied on Russian officials as the information sources, and this correlation is evident even during natural disasters. To be more specific, it was dominated by the two most influential state officials – Medvedev (predominantly) and Putin. The significance of the state's influence over media coverage is also supported by the observation that the state's position (especially of its leaders) has rarely been questioned by any newspaper. Even during the scarce coverage of the Kyoto Conference and especially during the Copenhagen Conference, Russia was presented as an environmental leader and a saviour of the negotiating process.

Overall, after conducting this media analysis, it can be concluded that there is a correlation between state policy and media policy on the subject of climate change. But other factors cannot be disregarded. For instance, coverage during the heatwave, even though it was not as high as during the Copenhagen Conference, was still quite significant, and anomalous weather conditions did bring additional attention to the problem. On the other side, as the discourse analysis demonstrates, and as mentioned above, the coverage during this time provided more opportunities for climate sceptics (which is rather strange considering that the consequences of climate change were demonstrated) and more space for sensationalist ideas about climate change. In contrast, when politicians became the main source of information, and when the state's position was clearly articulated, the coverage followed that lead. It became more supportive of the climate change thesis and less sensationalist, whilst the climate change mitigation process was perceived as a subject of greater importance and as a 'win-win' situation for Russia.

With that said, the comparison of media coverage in 1997 and 2009 shows that in the earlier period, when the government was resistant to taking action on carbon mitigation, the media reaction was not to primarily cover climate change from a hostile or sceptical perspective, but simply not to cover it at all. Even though the increase in coverage between the Kyoto and Copenhagen Conferences correlated with the rise of climate change as an important issue on the state's agenda, if the data collected are compared with those of other countries (see Chapter 5), coverage of all three events still shows the insignificance of environmental issues in Russia. Within five national newspapers during two months around some major events related to the climate change topic, the accumulated amount of articles did not exceed 127 articles per event. The next chapter explores this finding and argues that in the Russian case, the omission of the climate change topic is a greater problem than biased coverage of it could be.

Concluding remarks

One of the main messages of Herman and Chomsky's (1994 [1988]) analysis of the political economy of the American mass media states that the mass media are an actor in the free market economy and function according to its laws, by satisfying demands of owners, advertisers and information sources due to their economic dependence. Our media analysis has demonstrated that whilst newspapers can be in private hands, the state can still shape their coverage. This was especially evident through the analysis of the 'information sources' filters, which showed that Russian state officials act as a dominant informational source on the subject matter, and whenever they enter the discussion there is hardly any journalistic critique following their statements, which ultimately makes the Russian state the most authoritative newsmaker on this subject.

Furthermore, the application of the 'dominant ideology' filter (that the media will follow the lead of the strong Russian state) has shown that when the state slightly changed its stance on the problem and finally publicly acknowledged the anthropogenic character of climate change and Russia's commitments to GHG emissions reduction goals, the climate change topic entered the media discourse with the overarching message of Russia's leading position in the international negotiation process and a 'win-win' situation for the state. Besides the biased coverage of climate change in Russian newspapers and the absence of almost any critique of the elites' position on the problem, the preceding analysis has revealed the clear omission of the climate change topic from the Russian media discourse. Following the Russian position on climate change or not questioning its stance, the Russian media fail to create a full discussion of one of the most important and controversial environmental issues of our time.

Notes

1 Sarah Oates (2013, p. 6) states that 'according to the measurement by the World Telecommunications/ICT Indicators Database, 43 percent of the Russian population was online by March 31, 2011. The organisation reported an increase in usage of 1,826 percent between 2000 and 2010'.
2 Another way of comparing the papers' popularity and reach is to compare their circulations, where again *Argumenty i fakty* and *Komsomol'skaya pravda* take the leading roles. For example, *AiF* is the world's 17th largest newspaper (Newspapers24.com 2011), with a 2.3 million circulation, whilst *KP* daily publishes 655,000 newspapers and its Friday issue comes up to 3 million. Even though the other newspapers' circulations are much more modest, they are still considered to be *mass* newspapers.
3 'Left' and 'right' are quite ambiguous in the Russian context. During the Soviet Union the government identified itself as left wing 'by its [USSR's] leaders' hostility to "rightist reactionary" regimes abroad' (Evans and Whitefield 1998, p. 1024). During *perestroika*, 'right' and 'left' acquired absolute opposite meanings – 'the "left" came to denote the free market democrats and liberals, and the "right" the devotees of socialism and the communist system' (ibid). In the modern Russia, these terms were inverted once again, and for the purpose of this research, it is understood that newspapers that characterise themselves as 'left' are supporters of communist ideals, whilst 'right-wing' newspapers advocate the development of capitalism and liberalism, but at this moment the 'left' has become the opposition rather than the ruling party.

4 This is the so-called Lenin principle, which originates from his idea of how media organs function. Besides journalists working and writing for the newspaper, there also should be a network of regular people contributing to the issues who are not particularly educated to work in the newspaper but who share similar political views and have a 'grasp from the field' (more on this in Strovskiy 2011).

5 It is interesting that the readership age profiles seem to show that the print media are consumed mainly by a considerably older audience (typically 45+ for *KP*, 55+ for *RG*, and 65+ for *Izvestiya*, whilst 61 percent of the readers of *SR* are retired). One public opinion poll (FOM 2008) shows that there is only a small difference between respondents' attitude towards climate change problems depending on their age, such as in the 18–35 age group, 70 percent of people consider climate change to be an important problem, those 36–54 years old are 72 percent, and those over 55 years old are 65 percent . Based on this information, the argument that the print media orient their coverage towards their reader's interests is not very convincing; however, it needs further study.

6 The use of different databases raised a concern for the comparability of the researched data; however, it was noticed that both databases use the same principle of collecting and exporting articles. The controlled search was conducted where the same newspaper (which was on both databases) was searched with the same keywords and the results were identical.

7 Medvedev was advised not to go to the conference, and nobody knew if he would go until the last moment, so NGOs tried to do their best to influence him to take part in it (environmental activist, interview 27, Moscow, 22 July 2011).

8 Given the high number of observations (more than 3,000 articles per newspaper), our statistical power, i.e., our ability to discern population differences from the sample, is remarkably high. Therefore, the lack of statistical significance in most of the tests is unlikely to be due to a type II error.

9 'Corpus linguistics encompasses the compilation and analysis of collections of spoken and written texts as the source of evidence for describing the nature, structure, and use of languages. This work typically brings a quantitative dimension to the description of languages by including information on the probability with which linguistic items or processes occur in particular contexts' (Kennedy 2002, p. 2816).

10 The nonverbal message components will not be analysed due to technical reasons. It suggests the study of photographs, page layout, font size and type; however, these research data were accessed through computer databases and contained only the text and not the form of the articles.

11 Even though each of the selected articles was carefully studied through the methodology proposed by Mautner (2008), due to the word limit in the chapter, only the most frequent and prominent characteristics of the analysed texts are presented.

12 Since more articles were available for events of 2009 and 2010, the analysis will be divided into subcategories of different newspapers, unlike in the case of the Kyoto Conference, where such separation would be rather pointless.

13 Interestingly, both *Izvestiya* and *Kommersant*, with a two-day difference, published almost the same title – as mentioned, *Kommersant* produced an article with the title 'Medvedev Did Not Play a Role' ('*Dmitry Medvedev Pogody ne Sdelal*' by Granik 2009a), whilst *Izvestiya* produced one titled 'Copenhagen Did Not Play a Role' ('*Kopengagen Pogody ne Sdelal*' by Farizova (2009a). Literal translation of this expression means that somebody or something did not 'fix the weather' (which obviously correlates with the climate change topic). The actual meaning tells us about the unimportance of the actor or the inability to achieve something. In spite of the disparaging attitude towards the head of state in the *Kommersant* publication's title, both articles carried very similar messages. *Izvestiya* highlighted that about 'the only leader who announced his country's commitments on carbon emissions regardless of the conference outcome was Dmitry Medvedev' and *Kommersant* also wrote about Medvedev's 'clear position on climate policy' and the 'pacificator' character of his presentation.

14 One might argue that 'adequate' coverage of climate change is a vague concept. For example, there is a difference between improving media coverage of climate change by making it more sophisticated and by making it more sympathetic to taking action to mitigate climate change. Even though for the environmental activists and supporters of the active climate change mitigation policy, the latter would be the most desirable outcome, talking about the problem in terms of freedom of speech and Russian media becoming something different than just a propaganda tool, blind commitment to either side can be considered a step back for the development of democratic media.

15 Gruza and Ran'kova (2011) argue that in the case of the Russian heatwave, the main reason was a 'slow moving anti-cyclone'; however, global warming contributed to this disastrous weather event by increasing the created anomalous high temperature by 2–3 degrees.

Bibliography

Atlas SMI (2011), www.mediageo.ru, date accessed 20/06/2011.

BBC News (2008) 'The Press in Russia', *BBC News*, http://news.bbc.co.uk/1/hi/world/europe/4315129.stm, date accessed 14/09/2011.

Carter, R. and Simpson, P. (1989) *Language, Discourse and Literature. An Introductory Reader in Discourse Stylistics*, London: Routledge.

Carvalho, A. (2008) 'Media(ted) Discourse and Society', *Journalism Studies*, 9/2: 161–177.

Castro, F. (2009a) 'Pravda o Tom, Chto Proizoshlo na Sammite', *Sovetskaya Rossiya*, 24 December.

Castro, F. (2009b) 'Chas Istiny', *Sovetskaya Rossiya*, 22 December.

Castro, F. (2010) 'Velikan v Semimil'nykh Sapogakh', *Sovetskaya Rossiya*, 87, 17 August.

Chizhikov, M. (1997) 'Cherez 20 Let Nas Zhdet Vubor: Zadokhnut'sya ili Zakhlebnut'sya', *Komsomol'skaya pravda*, 5 November.

Elkov, I. (2009) 'Klimat-Kontrol', *Rossiyskaya gazeta*, 17 December.

Elkov, I. (2010) '40 v Teni', *Rossiyskaya gazeta*, 5 August.

Evans, G. and Whitefield, S. (1998) 'The Evolution of Left and Right in Post-Soviet Russia', *Europe-Asia Studies*, 50/6: 1023–1042.

Farizova, S. (2009a) 'Kopengagen Pogody ne Sdelal', *Izvestiya*, 21 December.

Farizova, S. (2009b) 'U Klimata Poyavilsya Privkus Deneg', *Izvestiya*, 16 December.

FOM (1997) 'Moskovskaya Auditoriya "Sovetskoy Rossii"', http://bd.fom.ru/report/cat/smi/smi_print/newspapers/of19972005, date accessed 15/01/2012.

FOM (2007) 'SMI: Predpochetaemye Kanaly Informatsii', http://bd.fom.ru/report/cat/smi/smi_rei/d073121, date accessed 15/06/2010.

FOM (2008) 'Ekologicheskaya Situatsiya v Massovom Soznanii Rossiyan', www.fom.ru, date accessed 20/09/2011.

FOM (2013) 'Istochniki Informatsii i Telepredpochteniya Rossiyan', http://soc.fom.ru/obshchestvo/10938, date accessed 2/10/2013.

Gamov, A. (2010) 'Vladimir Putin – Merzlovedam: "Kogda Podtopit ili Podmorozit – Predupredite, chtoby Znat' Kyda Bezhat" . . .', *Komsomol'skaya pravda*, 24 August.

Golovnin, V. (1997) 'Spory o Klimate u "Sada Kamney"', *Izvestiya*, 3 December.

Gorbachev, M. (2009) 'Kto Povyshaet Gradus?', *Rossiyskaya gazeta*, 8 December.

Granik, I. (2009a) 'Dmitry Medvedev Pogody ne Sdelal', *Kommersant*, 19 December.

Granik, I. (2009b) 'Global'noe Poteplenie dlya Rossii Ob"yavleno Rentabel'nym', *Kommersant*, 15 December.

Gruza, G. and Ran'kova, E. (2011) 'Otsenka Vozmozhnosti Vklada Global'nogo Potepleniya v Genezis Ekstremal'no Zharkikh Letnikh Sezonov na Evropeyskoy Territorii RF', *Izmenenie Klimata*, 30: 13.

Herman, E. and Chomsky, N. (1994 [1988]) *Manufacturing Consent. The Political Economy of the Mass Media*, London: Vintage.

Izvestiya (1997) 'Rossiya Mozhet Zarabotat' na Poteplenii Klimata', *Izvestiya*, 16 December.

Izvestiya (2010) 'Davlenie Prirody', *Izvestiya*, 9 July.

Kabannikov, A. and Potapov, P. (1997) 'Yapontsev Smoet, Irlandtsy Peresyadut na Oleney, Yanki Uidut v Lesa', *Komsomol'skaya pravda*, 18 December.

Kennedy, G. (2002) 'Corpus Linguistic', *International Encyclopedia of the Social and Behavioral Sciences*, www.sciencedirect.com/science/article/pii/B0080430767030564, date accessed 10/01/2012.

Khanzhin, V. and Khromov, V. (2010) 'Ni Xitru, ni Gorazdu Suda ne Minuti', *Sovetskaya Rossiya*, 85, 12 August.

Kharnas, A. (2009) 'Teploe Mesto', *Kommersant*, 7 December.

Kommersant website (2011), www.kommersant.ru, date accessed 19/06/2011.

Komsomol'skaya pravda (2010) 'Zhara-2010', *Komsomol'skaya pravda*, 12 August.

Kovyneva, T. and Moiseenko, A. (2009) 'Russkie Khakery Razoblachili Mif o Global'nom Poteplenii?', *Komsomol'skaya pravda*, 23 November.

Kramich, G. (2010) 'Toplivnaya Bomba pod Shaturoy', *Sovetskaya Rossiya*, 86, 14 August.

Krivyakina, E. (2009) 'Dmitry Medvedev – o Global'nom Poteplenii: "My i Tak Vyigraem, i Po-drugomu"', *Komsomol'skaya pravda*, 15 December.

Kuzina, S. (2010) 'Zhara v Rossii – Rezul'tat Ispytaniya Klimaticheskogo Oruzhiya?', *Komsomol'skaya pravda*, 29 July.

March, L. (2009) 'Managing Opposition in a Hybrid Regime: Just Russia and Parastatal Opposition', *Slavik Review*, 68/3: 504–527.

Mautner, G. (2008) 'Analysing Newspapers, Magazines and Other Print Media'. In: R. Wodak and M. Krzyzanowski (eds.) *Qualitative Discourse Analysis in the Social Sciences*, Basingstoke: Palgrave Macmillan: 30–53.

Media Atlas (2011), www.media-atlas.ru, date accessed 16/06/2011.

Medvedev, Y. (2010) 'Gradusom Vyshe', *Rossiyskaya gazeta*, 17 August.

Merinov, S. (2009) 'Nagruzka na Klimat', *Rossiyskaya gazeta*, 18 December.

Mironov, N. (2010) 'V Moskve-Reke Poyavilis'; Meduzy', *Komsomol'skaya pravda*, 5 August.

Moiseenko, A. (2009) 'Film-Katastrofa "Poslezavtra" Mozhet Voplotit'sya v Real'nost', *Komsomol'skaya pravda*, 17 November.

Motskobili, I. (1997) 'Rossiya Priznana Nevinovnoy v Global'nom Poteplenii', *Kommersant*, 19 December.

Motskobili, I. and Maksimenko, E. (1997) 'XXI Vek Proydet v Atmosfere Nechelovecheskoy Teploty', *Kommersant*, 29 November.

Newspaper24.com (2011) www.newspapers24.com/largest-newspapers.html, date accessed 25/10/2011.

Oates, S. (2013) *Revolution Stalled: The Political Limits of the Internet in the Post-Soviet Sphere*, Oxford: Oxford University Press.

Obraztsov, P. (2010a) 'Beloe Solntse Rossii', *Izvestiya*, 19 July.

Obraztsov, P. (2010b) 'Klimat Massovogo Porazheniya', *Izvestiya*, 13 August.

Online Gazeta (2011), 'Gazeta Kommersant', www.onlinegazeta.info, date accessed 20/06/2011.

Pertsovskiy, O. (2010) 'I Stanet Drugom Vrag', *Rossiyskaya gazeta*, 28 August.

Petrov, V. (2009) 'Antiparnikovyy Effekt', *Rossiyskaya gazeta*, 15 December.

Popov, E. (1997) 'A Klinton Puskaet "Solnechnye Zaychiki"!', *Sovetskaya Rossiya*, 4 December.

Rossiyskaya gazeta (2010) 'K Chemy Gotovit'sya', 17 August.

Rossiyskaya Gazeta website (2011), www.rg.ru, date accessed 16/06/2011.

Sakwa, R. (2008) *Russian Politics and Society*, Oxon: Routledge.

Sakwa, R. (2011) 'The Future of Russian Democracy', *Government and Opposition: An International Journal of Comparative Politics*, 46/4: 517–537.

Sapozhkov, O. and Butrin, D (2009) 'Britanskie uchenye Nedootsenili Russkiy Klimat', *Kommersant*, 16 December 2012.

Savinykh, A. (2010) 'Prem'er Provel Klimat-Kontrol', *Izvestiya*, 24 August.

Sborov, A. (2010) 'Leto Sprosit Strogo', *Kommersant*, 18 August.

Shapovalov, A. (2009) 'Protokol Upushchenykh Vozmozhnostey', *Kommersant*, 7 December.

Shapovalov, A. (2010) 'Parnikovye Gazy Popali pod Razdachu', *Kommersant*, 28 July.

Shmeleva, E. (2009) 'Na Chetvert' Men'she', *Rossiyskaya gazeta*, 8 December.

Shmeleva, E. (2010) 'Klimat bez Zhertv', *Rossiyskaya gazeta*, 30 July.

Shvarts, I. (1997) 'Kraynikh Net', *Kommersant*, 29 November.

Smirnova, Y. (2009) 'Global'noe Poteplenie – Vydumka Klimatologov?', *Komsomol'skaya pravda*, 3 December.

Smirnova, Y. (2010) 'Kogda Zakonchitsya Anomal'naya Zhara v Stolitse?', *Komsomol'skaya pravda*, 21 July.

Smol'yakova, T. (2010) 'Prognoz Ministra Trutneva', *Rossiyskaya gazeta*, 24 August.

Sokolov, M. (2009) 'Osnovy Naturfilosofii', *Izvestiya*, 15 December.

Strovskiy, D. (2011) *Otechestvennaya Zhurnalistika Noveyshego Perioda*, Moscow: Uniti-Dana.

van Dijk, T. (1991) 'The Interdisciplinary Study of News as Discourse'. In: K. B. Jensen and N. Jankowski (eds.) *A Handbook of Qualitative Methodologies for Mass Communication Research*, London: Routledge: 108–120.

van Dijk, T. (2001) 'Critical Discourse Analysis'. In: D. Tannen, D. Schiffrin and H. Hamilton (eds.) *Handbook of Discourse Analysis*, Oxford: Blackwell: 352–371.

van Dijk, T. (2011) *Discourse Studies: A Multidisciplinary Introduction – Volume One*, London: Sage.

Zyuganov, G. (2010) 'Politika Katastrof', *Sovetskaya Rossiya*, 92, 28 August.

5 Mediating climate change in Russia

Passing through the barriers

Concluding the analysis of media coverage of climate change in Russia, it is claimed here that media coverage on the issue stays within the broad politico-economic framework which is influenced and controlled by the elites and that the media hardly ever challenge the elites' position on the problem. The findings of this research show that with regard to climate change, Russian journalists do not face open forms of censorship or state orders, but the whole politico-economic system of the country works in a way that encourages coverage to stay within certain boundaries. In summary it could be stated that in the case of climate change, the state seems not to be a main 'client', but rather a main newsmaker that takes the lead. As discussed in more detail below, after the Russian government changed its position on the climate change issue, the Russian media turned their attention towards the problem, which was signified by an increase in their coverage of climate change and the dominance of the Russian state officials as sources of information.

This chapter returns to the debate (see Chapter 1) on the influence of micro-processes in the media coverage of climate change (such as the specifications of the topic and the role of journalistic norms). During the interviews conducted for this project, journalists noted that interest in climate change was often stimulated by natural disasters or abnormal temperatures (which was also confirmed by the media analysis data), and on the contrary the lack of interest in some cases could be explained by the complexity of the scientific data and abstract nature of the problem.

Furthermore, in the Russian case, silence on the climate change problem speaks louder than any biased or unbiased coverage. I argue that the omission of the climate change topic demonstrates what Lukes (2005) calls the 'third dimension of power', which the Russian state has exercised. Finally, based on the conducted research, at the end of this chapter it is possible to suggest a number of ways in which the climate change issue in Russia can be popularised and how adequate discussion can develop.

The media and power, or the powerful media?

As Oates (2007) rightfully notes, 'social scientists remain unsure as to whether the media tend to lead political change or (more cynically) if they merely reinforce the consensus of the political victors' (p. 1279). In the Soviet Union (especially at its beginning), 'the media had a particularly important role for the Soviet leadership in the creation of a fully communist society' (White and Oates 2003, p. 32). Even though the media were considered a powerful 'propaganda tool', with total control and institutionalised censorship, they were a factor rather than an actor in the state's communication strategy. The distribution of power was quite straightforward, from the top down, with the state dictating to the media what to do and the media helping the state to mobilise and 'organise the masses'.

More recently, during the period of *perestroika* and especially after the collapse of the Soviet Union and the creation of the new state (with a supposedly democratic regime), many (including journalists themselves) saw an opportunity for the Russian media to become the fourth estate. Furthermore, media would be capable of altering the political regime, to bring attention to the problems and place them on the agenda and actually be able to influence the outcomes of the elections. In this sense the media would become a 'watchdog' for democracy, and their main role would be 'to act as a check on the state [, . . .] monitor the full range of state activity, and fearlessly expose abuse of official authority' (Curran 2002, p. 217). Indeed, as Ivan Zassoursky (2004) puts it, 'the press in the early 1990s genuinely perceived itself as a "fourth estate", that is, as one of the governing institutions wielding enormous influence in society' (p. 57). It could be argued that for a few years this was the case, and one of the most famous examples of the Russian media being an active actor in the political process is when they organised an influential campaign against the first military action in the Chechen Republic (see Chapter 2). This 'golden age' of Russian media did not last very long and soon it turned into a neo-Soviet media system (Oates 2007), where newly acquired freedom of speech in many cases remained just a formality and additional constraints dictated by the free market were added.

The analysis of the media system in Russia demonstrates that if we break this system into elements, it becomes clear that all of them are dominated by the most powerful actor in Russia – the state. The conducted research showed that the state in one way or another owns the most important media outlets. It has close connections with the advertising market. It contributes to the reaction on media activity in some form of censorship. It often becomes the dominant information source and it creates a certain political or ideological regime in the country which some might call political capitalism, managed democracy or a dual state. In the study of media coverage of climate change through the series of interviews and the content and discourse analysis of the selected newspapers, the hypothesis of the state being a dominant independent variable has been proven to be correct.

The quantity of the articles changed significantly after the change in the state's climate policy (which was demonstrated through the comparison of coverage of the Kyoto and Copenhagen Conferences). The coverage provoked by the official events such as the Copenhagen Conference and acceptance of the Climate Doctrine resulted in almost twice as many articles as the natural disaster event (the heatwave of the summer of 2010). Since the analysed newspapers were selected according to the logic that they represent different types of ownership structures, the conducted research also has shown that the newspapers which are owned by the state quite predictably follow the state's agenda; however, the more interesting and unexpected results that were achieved through this analysis show that the newspapers owned by big companies or oligarchs are not so different in their media policy to the state-owned ones.

An interesting case study was presented through the analysis of the oppositional newspaper *Sovetskaya Rossiya*, which claims to be independent. In terms of how many articles each newspaper publishes per issue and how many articles were devoted to climate change, *Sovetskaya Rossiya* talked about climate change quite frequently, and every time, it accused the capitalist system (often simply the United States) for its destructive force and for global environmental degradation. At the same time, *Sovetskaya Rossiya* was not that different from the other studied newspapers, in the sense that it also did not question Russia's climate change policy and its contribution to the world's level of GHG emissions.

No presentation of climate change versus misrepresentation

In the first chapter of this book, the importance of the mass media in the process of mitigating or adapting to climate change consequences was highlighted. As Kokhanova (2007) states: 'before we will be able to mitigate or solve any kind of global environmental problems, firstly, we need to define those problems – an exchange of information should happen' (p. 19). The same idea is shared by Anders Hansen (2010), who argues that the media need to explain to people what the environmental problems are, especially in the case of climate change – 'whatever "symptoms" of climate change that we see around us, they are of course only just that because we have been told that this is what they are, manifestations of climate change' (ibid, p. 170). As has been discussed before (see Chapter 1), climate change is characterised as an 'unobtrusive issue' which cannot be noticed and understood without specialised knowledge of the subject or without it being 'translated' and 'broadcast' by the media.

The reluctance of the Russian media to cover climate change was picked up during the interviews. For example, an NGO representative noticed that it could be considered a positive sign that now media have started to mention climate change, whilst a few years ago even this was not the case (interview 19, Moscow, 27 July 2011). Another environmental activist supports this position by stating that in Russia 'mass media cover climate change awfully, but before they did it even worse' (interview 31, Moscow, 27 July 2011).

A study of the worldwide collaborative research network 'MediaClimate' (Eide et al. 2010) managed a comparative study of the media coverage of the Bali Summit (December 2007) and the Copenhagen Conference (December 2009) in 13 and 19 countries respectively; in both cases Russia ended up with the lowest amount of articles devoted to the topic.[1] For instance, during the conference in Bali, there were only 13 articles published in two Russian newspapers, (*Kommersant* and *Moskovsky Komsomolets*). For the Copenhagen Conference, the number was slightly higher, and Russian media outlets studied together managed to publish 32 articles, whilst the top positions were occupied by Denmark (the host country) with 710 pieces, Bangladesh with 317, Norway with 264 and Canada with 262. The countries closest to Russia in their coverage rate were El Salvador with 55 and Chile with 48, which was still at least 16 articles more than in the Russian case. Furthermore, only 17 articles in the analysed Russian newspapers were directly devoted to the Copenhagen Conference, whilst the other 15 just mentioned it in the discussion of other topics, mostly in the context of the new strategic weapons agreement between the United States and Russia (negotiations about which were happening simultaneously) (Yagodin 2010).

The analysis of the newspapers' coverage of climate change in Russia conducted for this research has confirmed these findings. As has been demonstrated, even within five national newspapers during two months around some major events related to the climate change topic, the total number of articles did not exceed 127 per event. The number of articles which were specifically devoted to climate change (rather just mentioning it in a non-related context) was even less. During two months around the Kyoto Conference in 1997, five newspapers managed to produce only 9 articles discussing climate change; during the Copenhagen Conference in 2009, it was 68 articles, and during the heatwave events of summer 2010, only 30 articles were written on climate change.

For the purpose of comparison, a similar search was conducted, through the electronic database Nexis, of UK newspapers. The result showed that in the five selected newspapers (*The Independent, The Guardian, The Times, The Observer* and *The Daily Mirror/Sunday Mirror*), 337 articles devoted to climate change were published during the Kyoto Conference and 1,744 articles were published during the Copenhagen Conference. To get comparative data from a country which did not join the Kyoto Protocol, the *New York Times* and the *Washington Post* were searched. During the Kyoto Conference, they respectively published 133 and 112 articles. During the Copenhagen Conference, the *New York Times* produced 291 articles and the *Washington Post* 260. One might argue that the absolute number of articles on climate change might lead to distorted conclusions (since in various countries, newspapers can have different approaches to news production – printing less or more news per issue). In the extensive comparative study, Schmidt et al. (2013) took into account this consideration and next to the absolute number of articles have also provided a percentage of news articles dedicated to (or mentioning) climate change to the overall number of publications in the respective media outlets. Once again Russia demonstrated one of the most modest results, where out of 27 researched countries it had the third smallest result, leaving behind only Jordan and Yemen.

So, one of the factors characterising the media policy on climate change in Russia which has been confirmed by this study is the relative omission of information on this controversial problem. Many scholars struggle to find explanations of why journalists write about climate change in a sensational manner, why they devote the same space and time to the arguments supported by climate sceptics as to the arguments supported by the dominant majority of scientists and so on. In the Russian case, one might say the problem is more complex – why this debate on climate change has not even entered the public discourse in any serious way.

One of the possible explanations behind the limited coverage of climate change which Russian journalists have offered themselves (various interviews) could be seen in the low public interest towards the problem. Whilst the significance of the public's influence over the media production process is a debatable question, in the Russian case of climate change coverage, the correlation between the low coverage of the problem and the low level of awareness of climate change among the general population is indeed quite striking.

The problem was demonstrated by the Public Opinion Fund (FOM 2008), which conducted an opinion poll where the respondents were asked to choose not more than 5 out of 25 options of the problems they are most concerned with. Whilst the most popular answers were 'inflation, price increase', 'high housing prices' and 'expensive medical care', 'environmental problems' took 19th place, leaving behind only problems connected with the immigration situation, public transport and delays with pay days (the last three places were taken by marginal answers falling into categories 'other', 'do not have any problems' and 'do not know'). What is interesting is that when people were directly asked whether they were concerned by the environmental situation and if they thought global warming was an important problem, 78 and 70 percent (respectively) answered affirmatively. Another study was conducted by Greenpeace Russia in 1999 which aimed to find out the public's attitude towards the NGO and charities in general and it showed that people did not mind supporting 'actions to protect the environment near their home or neighbourhood; the actions they were least likely to support were "pressuring the authorities and business for the goal of resolving environmental problems" and "the battle against global climate change"' (Henry 2010, p. 198, citing Greenpeace Russia 1999). The Yuri Levada Analytical Centre in 2014 demonstrated results that were slightly more 'positive'. Respondents to its recent opinion poll dedicated to environmental problems and environmental security were asked to select environmental problems which worry them the most. The problem of climate change came in 6th (out of 14 possible positions), with 23 percent of interviewees considering it one of the most worrying environmental problems. According to the available data, it can be seen that this perception of climate change has been more or less constant over the last four years (24 percent in 2011, 22 percent in 2012 and 21 percent in 2013). Hence, whilst climate change is far from being perceived as one of the biggest environmental threats (this niche has been occupied by water or air pollution), at the same time it did 'outrun' a number of other environmental problems.

A more specific study dedicated to climate change was conducted in June 2013 by the initiative of the Interagency Working Group on Climate Change and Sustainable Development, which showed that 54 percent of interviewees knew about climate change and 36 percent had heard something about this issue (President of Russia website 2013b). The poll also showed that 33 percent of respondents believed in the anthropogenic cause and 42 percent equally blamed humans and natural processes. A slight majority of people (53 percent) would support the introduction of economic stimulus for decreasing GHG emissions, whilst 41.4 percent were ready personally to do something to help the fight with climate change (40.6 percent were not prepared for this step). These results were interpreted as being somewhat positive (RSEU 2013), since they demonstrate quite high levels of climate change awareness in Russia; however, as demonstrated below, when Russian public opinion on climate change is studied in comparison with other countries, the results are very negative.

The World Bank (2010) commissioned an opinion poll which aimed to determine public attitudes across the globe particularly towards climate change. The research was conducted in 16 countries: Bangladesh, Brazil, China, Egypt, France, India, Indonesia, Iran, Japan, Kenya, Mexico, Russia, Senegal, Turkey, the United States and Vietnam. Even though the summary of the results states that overall people in all the studied countries demonstrated a high level of concern about climate change, Russia in almost all question categories occupied one of the last places, showing the lowest level of concern amongst its citizens. For instance, 30 percent of respondents in Russia considered climate change to be a 'very serious' problem, whilst the opinion poll average was 60 percent. Only 18 percent of Russian respondents 'strongly agreed' that 'dealing with the problem of climate change should be given priority, even if it caused slower economic growth and some loss of jobs', whilst the average among the 15 other countries was 35 percent. Russia had the least number of people who thought that the majority of scientists 'think the problem is urgent and enough is known for action', at 23 percent, whilst this multi-country poll showed that on average 51 percent of people agreed with this statement. Once again Russians are the most negative with regard to the question of whether their country had the 'responsibility to take steps to deal with climate change', 58 percent, whilst the world average showed that 87 percent of people thought their state should be responsible for dealing with the problem. These results echoed the opinion poll commissioned a year earlier in 2009 by the BBC World Service, which stated that out of 23 studied countries, Russia demonstrated one of the lowest results in referring to climate change as a 'very serious' problem (46 percent of respondents had selected this option whilst the world's average was 64 percent) (Mountford 2009).

Arguably, this rather low level of concern with environmental problems in general and climate change in particular can serve as a justification for the limited media coverage of climate change in Russia. As is discussed further on in this chapter, this is a questionable statement – is press coverage limited because readers are not interested, or are readers not interested because coverage is limited?

The role of media in forming public opinion and generating public interest is explored to a great extent in the studies devoted to the media's 'agenda-setting' capabilities. For example, McCombs and Shaw (1972, p. 176) argue that with the 'help' of mass media 'the readers learn not only about a given issue, but also how much importance to attach to that issue' (see also Carroll and McCombs 2003; McCombs 2004, 2005). At the same time, James Curran (2002), in his discussion of the 'limited media influence' (from the perspective of the liberal approach to media theories), questions the power of media over their audiences by stating that 'audiences selectively attend to, understand, evaluate and retain information from the media' (ibid, p. 132). So regardless of how well the media propaganda machine operates, people do not simply absorb without questioning everything the media drops upon them:

> This is because the public is not an empty vessel waiting to be filled by media propaganda[. . . .] Even when people are exposed to communications from the media on a topic they know nothing about, they have core beliefs and general orientations – 'interpretive schema' – which results in selective assimilation of information.
>
> (Curran 2002, p. 132)

In the case of climate change, such personal characteristics as social class, age, gender, level of education and even religious beliefs can alter people's perception of risks to the environment (Smith and Leiserowitz 2013, Zhao 2009). This said, considering the relatively significant influence of the Russian media over their audience, and the low level of general awareness of the problem, an increase in the level of coverage can lead to an increased understanding of the problem. As White and Oates state, 'the media are more trusted than any other social institution in contemporary Russia – more than the armed forces, the Church, political parties or government itself' (2003, p. 33). Hence, even though the public influence over media coverage of climate change is debatable, the possibility of media power increasing people's awareness of the problem in Russia is rather high. Nenashev (2010) states that in contrast to Western media, where objective information and impartial coverage are priorities, Russian journalism has always preferred commentary and analysis of events, which often include a direct appeal to solve the political or social problems or provide people an option of not only *what* to think about but also *how* to think.

Indeed, throughout the history of Russian media, regardless of the degree of their dependency on the state or other actors, they have always possessed the specific characteristic of not just being an informer but rather an educator. Interestingly enough, the Russian public is mostly content with the nurturing role of media. As White and Oates (2003) found in their extensive empirical research on the public's attitude towards media in Russia: 'many [respondents] thought it was simply irresponsible of the mass media to present information in a neutral way, without any kind of reference to wider moral or patriotic values' (ibid, p. 33). Another of White and Oates's interesting findings showed that 'Russians are often more distressed by the portrayal of violence and chaos on their television screens than by pro-government bias' (ibid). Even more shockingly, according to this research, the Russian people have a more positive attitude toward media coverage after the centralisation processes of Putin's regime compared with the era of media 'freedom' in the early 1990s (mostly due to the fact that media coverage of the time was also quite chaotic and devoted large amounts of space or time to topics with violent or sexual content).

Due to these concerns with media abusing these questionable topics, the public opinion poll conducted by VTsIOM showed that around 70 percent of respondents were in support of some kind of censorship over mass media (Tarusin and

Fedorov 2009). Interestingly enough, the same ideas were voiced by a number of interviewees in conversations about Russian media policy and climate change, where they said the problem is not in state policy:

> [T]he problem is in the media themselves, they like to shout about their freedom, but it comes down to the talk about scandals, so we need some kind of control from the civil society, from the expert community. I do not argue in support of control over the media (we already had it), but something has to change, they need to become more responsible.
>
> (NGO representative, interview, Moscow, 2011)

The public vision of the censorship of media activity comes down to a very simplistic concept – the media can do anything if they do no harm (by exposing too much of the above-mentioned controversial topics).

'Climate silence' and the state

The explanation behind the phenomenon of the media's reserved reporting of climate change was unfolded throughout the previous chapters, which have demonstrated the way Russian media operate and the correlation between media coverage of climate change and state policy towards it. Despite the recent modifications in the state's climate change policy, it is still a low-priority issue. For example, at the end of the 1990s (around the time of the Kyoto Conference), the Russian government was concerned more with the economic crisis of 1997 and Yeltsin's relations with strong political actors at the time – the oligarchs. The new chapter in Russia's political history and its influence on media has been studied in detail beforehand, but in general, Yeltsin's rule and the start of Putin's time were mostly characterised by an orientation towards economic problems and Chechnya, whilst all other issues (including environmental problems) were postponed until 'better times'.

The same attitude could be witnessed amongst the general public. As the deputy editor-in-chief of the Fund for Independent Radio, Elena Uporova, stated, 'we all know well that poverty goes along with the lack of interest in environmental topics as well as the fact that the power elite does not give signs that this [environment] is important' (presentation at the seminar 'Forest and Climate', Chemal, 14 August 2011). A similar view was expressed by an editor of the regional newspaper: 'our country was going through the long-term crisis of its political regime and it had to prioritise – in a hungry country nobody cares about the environment' (interview 16, Chemal, 14 August 2011). So, like the pieces of the same jigsaw picture, the public's, the state's and the media's diminution or underestimation of the climate change problem all contributed to the same outcome where the problem remains at the bottom of the hierarchy of needs and interests and does not get closer to its solution.

In order to break this pattern, this research project has tried to answer the question: who is leading whom in ignoring climate change? During the interviews, journalists often claimed that they did not write about climate change because the audience was not interested in it. Rather it is very difficult to make the problem look or sound relevant to people's everyday interests or needs and to give people a clear and straightforward answer about what climate change is and what it means for them. Indeed, this argument is consistent with the findings which were made by researchers studying problems of media coverage of climate change in various countries (see Chapter 1). However, this does not explain why, if Russian journalists face the same problems of public ignorance as their foreign colleagues, Russia is so far behind in the amount of articles written on this topic compared with other countries. American, British or Norwegian journalists facing the same problems in their professional activity manage to write about climate change up to ten times more often than their Russian colleagues. And why do Russian journalists not use the climate change topic as an opportunity to question politicians' performance on the matter (which indeed is quite questionable), as their foreign colleagues do? If this could be explained by the better economic situation in other countries (since often Russian journalists and policy makers state that environmental discourse will enter public space in Russia after key economic problems are solved), what about such countries as Chile and El Salvador?

These questions can be answered by an earlier stated argument of the predominant influence of the state policy on climate change. As an environmental activist who closely works on the climate change topic stated:

> Interest appears when something happens at the state level. For instance, when the Climate Doctrine was accepted, questions arose. The same happened during the announcement of the action plan for the doctrine. We [the NGO he represents] can clearly see now that journalists started to follow the climate problem, even though their interest is still very episodic (they only react on certain events) but at least it is not only based on sensationalism'.
>
> (interview 19, Moscow, 27 July 2011)

Prominent Russian scholar Oleg Yanitsky (2009, p. 759) also concludes that in the third phase[2] of the environmental debates, 'top state officials together with top media managers decided who would have access to the media.'

The relative omission of the topic could also be explained by the state's influence, or more precisely by the state's success in taking this topic 'off agenda' by exercising the 'less apparent face of power'. This thesis, introduced by Bachrach and Baratz in 1962, points out that even though it is commonly accepted that 'power is exercised when A participates in making the decision that affects B', they also argue that 'power is exercised when A devotes his energies to creating or reinforcing social and political values and institutional practices that limit the scope of political process to public consideration'. As the authors further quote Schattschneider's famous remark: 'some issues are organised into politics while others are organised out' (ibid, p. 949). Robert Babe (2005, p. 187) refers to this problem in his study of the media coverage of global warming in Canadian newspapers, where he states that 'the daily press, financed by advertisers and usually owned by multimedia organisations, will downplay the conflict between economic system and environment', which was shown both by the biased coverage and the omission of certain aspects of the problem. In the Russian case, 'the conflict between economic system and environment' should be 'replaced' by the 'conflict between the state interests and environment', then the conclusion of the omission of the 'inconvenient' topics would be very similarly reaffirmed.

However, whilst in Bachrach and Baratz's vision of power, the state's role in media coverage in Russia would suggest that state officials purposely remove controversial issues from the agenda, Steven Lukes's (2005, p. 25) 'three dimensional' extension of this power debate argues that power is exercised not only when certain decisions are consciously made or not made – 'the bias of the system can be mobilised, recreated and reinforced in ways that are neither consciously chosen nor the intended result of particular individuals' choices'. Lukes suggests various ways in which potentially controversial issues are 'kept out of politics, whether through the operation of social forces and institutional practices or through individuals' decisions' (ibid, p. 28). For instance, Lukes (2005, p. 144) refers to the Gramscian idea that ' "submission and intellectual subordination" could impede a subordinate class from following its "own conception of the world" '. Lukes further continues: 'Gramsci viewed civil society in the West as the site where consent is engineered, ensuring the cultural ascendancy of the ruling class and capitalism's stability'. Heyward (2007), in his revision of Lukes's three-dimensional view of power, notices that in the first two approaches to power, the 'agents can always identify and articulate their own interests', but 'if the third dimension of power was successfully exercised, even a slave might be content with his exploitation'. As Chomsky (1998, p. 187) argues:

> You cannot be a good propagandist unless it's in your bones[. . . .] So when people talk like this, you'll read liberal columnists in the *New York Times* very angrily saying, 'nobody tells me what to write. I write anything I feel like,' which is absolutely true.

In this respect, it is useful to refer to Lukes's explanation of the 'inactive power':

> [T]he features of agents that make them powerful include those that render activity unnecessary. If I can achieve the appropriate outcomes without having to act, because of the attitudes of others towards me or because of a favourable alignment of social relations and forces facilitating such outcomes, then my power is surely all the greater.
>
> (Lukes 2005, p. 79)

Hence, in this situation the powerful elites do not have to control every step of the journalists, they do not need to enforce sanctions, threaten correspondents or dictate the news agenda. Rather the topics will 'naturally' enter or leave the public discourse, such as happened with the abandonment of climate change issues in Russia – the topic was removed from the agenda and remains relatively unpopular due to the preoccupation of the elite (the state) and consequently the media with other issues. Evidently, this dependence on the state's attention to the problem makes climate change coverage extremely vulnerable, and as discussed below, the fears were raised that the recent modest burst of activity in communicating climate risks will again disappear from the agenda with change at the Russian executive level which might handicap the state's climate policy.

The role of personalities – will the situation get worse in Putin's third term?

The analysis of Russia's media and climate change policies has allowed us to conclude that in this case we witness the dominance of certain elite groups over media coverage of climate change. Specifically, the significance of the role played by Russia's main 'newsmakers' in the country – the state leaders – is clearly demonstrated. Like many others, one of the interviewed journalists stated: 'as soon as the president started to talk about it, everyone started. Yes, I think the peak of media activity [media coverage of climate change] does coincide with the position of the Kremlin' (interview 30, Moscow, 20 July 2011).

The specific characteristics of the Russian political system discussed earlier in this book confirmed that at the start of 'Putin's era' in 2000 the system once again was modified with the rapidly strengthening powers of the presidential post. After Putin's second term it became apparent that it was not about redistribution of powers towards a more presidential-focused type of political regime, but towards Putin himself (see more in Hanson 2010). As many have noticed, after Medvedev's succession to the post in 2008 Putin still maintained a significant amount of weight in Russian politics (arguably not proportionate to his post as prime minister at that time). With regard to this problem, Monaghan (2012, p. 2) argues that 'Medvedev was more liberal and more inclined to Russia's modernisation, yet was the weaker figure and without a political support base, whereas Putin was stronger, with a well-established support base, and was more focused on maintaining the status quo'.

The renewed interest in environmental problems at the state level has been associated with the start of the election campaign of the first vice–prime minister, Dmitry Medvedev, when he included the issues of environmental degradation and protection in his speeches (Bogdan et al. 2009). After Medvedev became president, multiple signs of Russia's more active climate policy followed (see Chapter 3). The influence of the personalities (heads of state) over the existing discourse around climate change problems was also demonstrated through the media analysis conducted for this research, which showed that the filter 'information source' was overall dominated by 'Russian official sources'[3] and in particular by two specific individuals – Putin and Medvedev (see Chapter 4). So, one of the logical questions which arises from the results is what will happen to climate change policy in general, and particularly after the presidential elections of 2012 and the second swap of offices between the two men, with Putin returning to the presidency (possibly for another twelve years).

One of the fears voiced during this research was concerned with Putin's 'special attitude' towards climate change issues, which with his comeback might lead to an age of stagnation in climate change policy or loss of the Russian media's interest in the topic, which has just started to develop. Andonova (2008, p. 491) argues that seeing changes in climate policy solely as a result of changes at the executive level of government would be 'an oversimplification of political reality'. Chapter 3 demonstrated that Medvedev's policy was moved not only by his striving for economic modernisation, but by the ideas which will remain relevant for any Russian leader. In order to test this hypothesis, a similar analysis (see more in Chapter 3) was conducted which included the study of official speeches and statements after Putin resumed his presidential position. Taking under consideration the fact that he has been in the post for less than one year (at the time of conducting this part of the research), this analysis suggests that in Putin's official discourse we can witness the reappearance of the same messages of *pragmatic environmentalism* rather than the previous (and very evident) ignoring of climate change issues. However, it is acknowledged that more extensive research should be conducted towards the end of his term.

From May 2012 (the start of Putin's current term) to April 2013, overall 13 texts with some reference to climate change were identified. This figure on its own shows that the climate change topic has remained relevant to the new administration. By comparison, during Medvedev's presidency, there were on average 18 texts per year mentioning climate change, with peaks coinciding with the Copenhagen Conference (27 texts identified in 2009 and only 5 in 2011) and acceptance of major documents concerning climate policy. However, unlike in Medvedev's case, during the first year of Putin's third term, there were no texts completely devoted to climate change or that discussed it at length – rather, they mentioned climate change among other items. Indeed, Putin's rhetoric on climate change differs from Medvedev's. Whilst Medvedev explicitly talked about Russia's position on climate issues, in Putin's statements we can observe only brief references to the problem. At the same time, it is also evident that the messages which Medvedev popularised in his official discourse have remained relevant for the new administration. Among the categories previously identified (Table 3.2), only two were discovered in the texts of Putin's presidency (Table 5.1); however,

Table 5.1 Percentage of Putin's speeches (2012–2013) by identified categories

Category	%	Examples of Quotations
Global cooperation	54	'Russia and Bangladesh negotiated to continue their co-operation on the issues of global climate change'.
		'The international community is facing an urgent need to find a way to effectively fight global challenges (such as climate change)'.
Economic benefits/ green economy	39	'We are convinced that economic development should not contradict the interests of environmental protection'.
		'The "Nord Stream" will work fully automatically under constant supervision of the control centre, without intermediate compression stations, which will reduce operating costs and reduce CO_2 emissions'.

they were among Medvedev's most popular themes – 'global cooperation' and 'economic benefits'.

Hence, the economic benefits which follow from the policies of energy efficiency stayed salient for the new Russian government. The messages of sustainable development and the green economy repeatedly enter presidential statements. Interestingly, one of the statements during the first year of Putin's presidency (latest) also discussed the negative consequences of climate change and in particular its effect on food security (however, as in the majority of the texts, no significant details are presented on this account). As the negative influence of climate change is almost certain to grow over the coming decades, this factor will force the government to abandon its policy of 'de-environmentalism' and think of ways to diminish the negative consequences of climate change.

Lastly, the majority of Medvedev's and Putin's speeches mentioning climate change discuss it within the context of global cooperation. As the president of the NGO Centre of Russian Environmental Policy, Vladimir Zakharov, said:

> [T]he change [swap between Medvedev and Putin] will play a certain role, but I think, it is now impossible for Russia to turn back in its climate policy (unless it drastically changes its political regimes and closes up to the West again), we will keep integrating into the world community. The pace might differ depending on circumstances, but overall the forecast is optimistic; we will keep paying more and more attention to environmental problems.
>
> (interview 29, Moscow, 21 July 2011)

An environmental activist (interview 19, Moscow, 27 July 2011) admitted that he believes that the situation is currently at such a stage that it cannot be ignored anymore. Furthermore, as he states, it is impossible for Russia to keep taking part in international negotiations (not necessarily connected with the climate change topic) and to claim that the problem still does not exist, 'considering that Russia is among the biggest GHG emitters – the world community will not leave us alone' (ibid). Correspondent John Harrison (radio 'Voice of Russia') said that he has not noticed the 'climb down' of climate change issues after Putin's return (Skype interview, 18 June 2012). Another influential figure in Russia's climate change policy, from a major environmental NGO, suggested the possibility of 'climate stagnation' in Russia and the decline of journalists' interest in the topic (considering that Putin again will not be treating the problem as a serious one and will be reproducing jokes such as the one on 'fur coats'). At the same time, he repeats the ideas expressed by other interviewees that it should not affect Russia's overall plan in reducing its GHG emissions, because this is connected with its technological development (interview 31, Moscow, 27 July 2011).

Reasons for optimism about Russia's climate policy can be seen in such 'deeds' as Putin's preservation of the president's climate change adviser's post, the development of laws and initiatives on energy efficiency and economic modernisation, as mentioned above, as well as the appointment (by Presidential

Decree 13/12/2012) of the Interagency Working Group on Climate Change and Sustainable Development. Furthermore, on 23 March 2013, for the first time, the Kremlin (the president's residence) and Red Square took part in the global 'Earth Hour', the main aim of which was to attract attention to the climate change problem by switching electricity off for one hour. The official announcement on the presidential website stated that the decision of Vladimir Putin to join this event 'is due to his traditional attention towards environmental problems, such as the declaration by Presidential Decree, 1157 (11/08/2012), that 2013 is the year of environmental protection in Russia' (President of Russia website 2013a). Towards the end of 2013, Vladimir Putin signed another presidential decree (N 752), 'On the reduction of GHG emissions', which suggested that the Russian government has to come up with measures to ensure that the state's GHG emissions will stay below 75 percent of its amount in 1990 (*Rossiyskaya gazeta* 2013).

Can media coverage of climate change in Russia be changed?

In the Russian case of media coverage of climate change it can be concluded that in order for the issue to become more evident, it should fall into the area of an interest of the state. Then it should comply with the interests of the economic elites: the financial organisations on which the media might depend (however, as discussed, the interests of economic elites should not contradict the interests of the state). As was mentioned on numerous occasions in Russia, these elite groups are extremely difficult to separate, or speaking more precisely, it is difficult to see the economic interests without the government's influence over them.

Further on, after the media pass these 'barriers' and reckon that the climate change topic is now in the interests of the 'main newsmakers' in the country and does not interfere with the major political and economic interests, then the other factors start to have an influence – for example, difficulties connected with understanding the scientific information, making information more interesting and relevant for the audience, finding the legitimate information sources (besides the official ones) and so on. Graphically these processes could be pictured as shown in Figure 5.1.

The foundation of the pyramid was explained earlier on in this book, but a few extra words need to be said about the top level. Though the collected data demonstrated the impact of the state's climate policy and position on these issues, in some cases it is not that straightforward. For example, the number of articles mentioning climate change-related issues during the non-politicised event of the heatwave in summer 2010 was still quite high, and a certain amount of government critique was presented on the pages of newspapers (for instance, with regard to the bureaucratic obstacles on the way to fulfilment of JI projects in Russia). During the interviews, quite often journalists stated that sometimes they did not write about climate change because of various reasons, such as the complexity of the topic, the difficulty of making the topic sound relevant to their readers, the desire

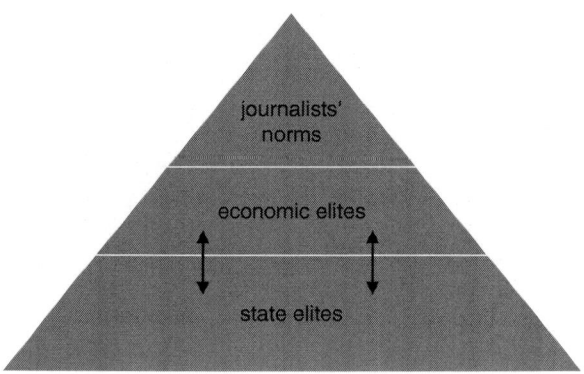

Figure 5.1 The three-stage model of media coverage of climate change in Russia

to write about 'hot topics' rather than the prolonged process of climate change and the readership's overall lack of interest in or ignorance of climate change issues.

The findings collected during the analysis of media coverage of climate change in Russia show that certain disagreements and variations in reporting on climate change can be witnessed. But overall they did stay within the greater consensus, characterised by the extremely limited attention to the problem among journalists and the absence of a sound critique of official Russian climate change policy. Even though in the end, in order to write about climate change, journalists do have to consider various micro-factors, this research project has shown that it is very unlikely that interest in the problem will be maintained if it falls out of the powerful elites' area of interests.

It is argued that media coverage of certain events might be altered if the elites change their position on the problem, or if the elites fall out among themselves, allowing various messages to enter the media space. In the Russian case, due to the significant dominance of the state over other actors that might be involved in climate policy, disagreement is less likely to happen. On the other hand, as was also demonstrated, climate policy not being the most important policy area for the state also has some benefits – for example, there is much less understanding of how the topic should or should not be covered, hence there is less possibility of control or, in the case of media activity, of censorship.

Based on the research conducted and the assumption that currently the idea of developing a greater discussion of climate change issues coincides with the state's more proactive climate policy, several conclusions were reached on what could be changed in order for Russia to bring the topic of climate change to the same level with other topics of national interest. The first two suggestions are concerned with the state and economic interests, and they were addressed due to their superiority; the other two suggestions are on the increasing role of the expert and activist communities in addressing the issues which concern the 'top' of the pyramid presented in Figure 5.1. That is to say, they address the problem of how the debate on climate change in Russia could become more sophisticated in the case where it *does not* contradict the elites' interests.

A more defined and visible state position on climate change

Experts from the Russian Regional Environmental Centre state that one of the main problems of Russia's climate change policy is the weak connections among scientists, politicians and the public, as well as low media coverage of climate change (Bogdan et al. 2009). In order to change the situation for the better, first of all, cooperation among all members involved should be established; however, due to the state's overriding influence, perhaps change in this area is of greater importance.

Russian climate change policy has indeed become more active in recent years. It is especially important that one of the most recent documents in this area – the comprehensive plan for implementation of the Climate Doctrine of the Russian Federation (2009) – includes such action points as dissemination of knowledge on energy conservation, energy efficiency and renewable energy as solutions to the problem of anthropogenic climate change. Another point of the plan prescribes providing public access to information on climate change and its influence on life (the organisation responsible for the first action is the Ministry of Economic Development; for the second, the Ministry of Natural Resources and Environmental Protection). So the necessity of communicating the state's policy on climate change to the media has been officially acknowledged, and the next step would be to establish and maintain the channels of communication between the state organisations and the media outlets.

For instance, one of these channels became the post of the president's adviser on climate change issues, which fortunately was preserved after Putin's return to office. The appointment of a prominent scientist to the post, Alexander Bedritsky, strengthened the ties between the state and science and resulted in a qualitative improvement of the president's speeches and remarks on the subject, which in turn improved communication with the media. The next step could be the establishment of a special state department working on this problem (like the Department of Energy and Climate Change in the UK), which would develop the idea of the recently created Interagency Working Group on Climate Change and Sustainable Development. Of course, quite obviously, attention to the problem will rise significantly if President Putin would become more proactive and explicit in his rhetoric on climate change issues (at least in the way Medvedev did by clearly articulating Russia's GHG emissions reduction goals and attending international conferences).

Perhaps, in the case of climate change, another way to solve the problem of media covering climate in a certain way (or not covering it) is to look at it from different angles and take it to the level where the coverage does not contradict the 'consent' amongst the elites – for instance, to talk about climate change on a personal level of how people can contribute to the fight against climate change, or on the contrary bring it to the international arena, and talk about climate change considering how it might improve or damage Russia's image and international relations.

For instance, Russia has a unique position in being part of key international organisations such as the G8 (which includes highly developed countries with vast interests in climate change negotiations)[4] but also belonging to industrialising countries' organisations such as BRICS (the actors with an ambivalent stance on climate change mitigation policy); and, finally, being one of the biggest energy exporters, it can relate to energy-rich developing countries (which are mostly sceptical or even hostile to the negotiations on GHG reduction goals) (Bagirov and Safonov 2010). So, in this sense Russia could be portrayed as an ambassador for the conflicting sides involved in the climate negotiation process. This is basically what Russian media has already been trying to do but in a very restricted way. Of course, all of these issues are also political and to some extent controversial. However, hypothetically, they allow media actors to present the topic in a less divisive way, where attention is removed from the state and the issues are discussed from a personal point of view or in a way that is beneficial for Russia.

In conclusion, in order to popularise the climate change topic in Russia, the media could focus not only on the powerful domestic actors but also at other levels of analysis. For example, Yanitsky (2009, p. 764) states that 'Russian society today needs a strategic dialogue with Europe', and he sees this as a way to bring the environmental debate in Russia to the next level: 'the dialogue should be taking place on continental as well as regional and local levels'.

Climate change as economic news

The next possible way to improve media coverage of climate change is strongly connected with the previous one, since it once again involves the state, but this time from the perspective of its close ties with the country's economic system. As Uporova states, 'editors-in-chief do not like social topics, if nothing particularly interesting is happening in there. That is why you [journalists] need to trick them – to present environmental information as, let's say, an economic one' (seminar presentation, Chemal, 13 August 2012). Moreover, many researchers argue that the common misperception among the people in charge of climate policy is that they treat the problem as purely environmental and keep it separate from economic issues. They do not see the dual direction of the climate mitigation programmes, which do not just allow for solving the problem of climate change but also stimulate energy efficiency programmes, development of renewable energy and so on (Bogdan et al. 2009).

This idea of Russia being able to gain a double benefit by following climate change mitigating policy through economic growth and achieving environmental development has been discussed in Chapter 3 as well as how this approach is becoming more pronounced in the statements of the state officials picked up by the media (see Chapter 4). Perhaps even more reinforcement should be made in this direction, since it seems to be the least controversial and contradicts the interests neither of the state and industry nor of the environmentalist and scientific communities. As mentioned in the previous chapters, at least over the next few decades Russia will be able to maintain its position as an exporter of oil and gas (its overall success depends on the geopolitical situation) with fossil fuels remaining the foundation for the world's economy, which will ensure a stable demand.

Therefore, it makes sense to reconsider the ways in which Russia can also offer its environmental services, and attract investments aimed at the decarbonisation of its economy. In this case the exploitation of the idea of Russia being a 'Great Ecological Power' or an 'environmental donor' (see Chapter 3) could be especially beneficial. Since the state policy in improving energy efficiency would tremendously contribute to the state's and the world's carbon emissions reduction goals, in this sense, Russia would really lead the way in climate mitigation policy (rather than keep referring to its involuntarily emissions' drop).

The NGO activist shared her experience of changing the approach to the problem of climate change:

> Through our [unfortunate] experience in working on climate change problems in Russia we realised that we should not be so direct with this topic, since it does not work. So, now we are working through the problem of food security[5] and we already can see that it is much easier to get to people [including journalists] through this topic, which is also affected by climate change, but is more tangible.
>
> (interview 27, Moscow, 22 July 2011)

So, food security would be another economic and social topic which is also connected with climate change but perhaps easier for the media and general public to comprehend and relate to. Another potential economic side of the climate change problem which also would be relatively easily popularised by mass media is carbon trading, for which Russia has great capacity, especially if it follows a policy of energy efficiency and development of the renewable energy sources as discussed earlier. 'Being realistic, I should admit that as soon as carbon trading starts developing in Russia, journalists will follow the money. For instance, look at the experience of the serious federal newspaper *Kommersant*: they usually write about climate change in terms of some economic problems.[6] I think most of the media outlets can adopt the same strategy' (Dobrovidova, correspondent at RIA Novosti, interview 9, Moscow, 20 July 2011). However, Russia's exit from 'Kyoto 2', the second round of the Kyoto Protocol environmental protection treaty, makes the fate of carbon trading in the country unclear (though there is some possibility of developing the national carbon trading market).

The 'economic framing' of climate change in Russia can also come in negative terms. For instance, Pralle (2009), in her analysis of how to keep climate change on the political agenda, recommends that climate change advocates emphasise 'human health impacts' and the 'cost of doing nothing'. The latter builds on the idea propagated by Nicholas Stern (2007, p. i), that 'the benefits of strong and early action far outweigh the economic costs of not acting'. As has been discussed beforehand, Russia is not exceptional in this regard, and its economy is already losing millions of rubles due to climate change consequences. The former statement similarly concentrates on the losses – however, it makes the problem personalised, giving specific information on how Russians will be directly influenced by this environmental issue. Apart from the human losses due to the extreme weather events, according to the Second Assessment Report of Roshydromet (2014) in the last decade we can witness an increasing number of cases of West Nile virus and tick-borne encephalitic diseases due to the climatic changes in Russia.

In summary, the economic 'card' could arguably be the strongest one in attracting more attention from the Russian media to climate change. As Zakharov states, 'now that economic growth is working on us [Russia], the richer we get, the more attention will be diverted to environmental problems' (interview 29, Moscow, 21 July 2011). Also, if the topic is approached from the position of potential benefits or elimination of losses, then it will not contradict the wishes of either the state or business elites.

Publicising Russian science

Until a few years ago, the Russian scientific community was creating additional barriers in the way of popularising the climate change mitigation policy. Mandrillon (2008) characterises that community's performance during the Kyoto ratification process as follows: 'scientists have not only failed to issue any warnings but, when their opinion has officially been sought, they have expressed opposition to the Kyoto Protocol' (p. 143). On the other hand, the positive change (both quantitatively and qualitatively)[7] of coverage of climate change issues in Russia is attributable to the shift in the role that scientists play in addressing this problem. This shift was signified by the appointment of the climate change adviser to the president in 2009, or by Russian scientists' contributions to the Intergovernmental Panel on Climate Change's Fourth Assessment Report on Climate Change in 2007, which included the most up-to-date information on climate change and its consequences in Russia (Bogdan et al. 2009).

Furthermore, climatologists have become more proactive in communicating their knowledge on the subject and making it more accessible for the wider public, including journalists. A successful example of scientists' attempts to share their information on climate change is the launch in April 2009 of a monthly electronic newsletter, *Izmenenie Klimata* ['Climate Change'] by Roshydromet. According to the editor of *Izmenenie Klimata*, Dr. Pavel Vargin (interview 25, April 2013), the idea to create the newsletter 'was in the air' for a while – 'to improve the communication of climate change risks is an acute problem for all countries including Russia'. Vargin notices that 'often you can see in the mass media pseudo-scientific discussion about climate, hence we try to publish opinion and comments of the most prominent scientists in the area of climate change from Roshydromet, the Russian Academy of Science and so on'. As the founders of the newsletter state, its main purpose is to communicate complicated messages about climate change to the broader public and raise general awareness and understanding of the problem. Anyone, whether a journalist or an ordinary person interested in climate change affairs, can subscribe to it for free. Currently the newsletter has 435 subscribers, including various academic organisations, NGOs, international organisations, foreign diplomatic missions in Russia and (particularly important for this research) the Russian central and regional mass media (Roshydromet 2012). During the fieldwork for this project, several respondents (journalists and representatives of NGOs) noted that this newsletter had been of great use to them and that they would like to see more initiatives like that coming from the scientific community.

Despite the increase in such positive practices by scientists, another obstacle to improving the media coverage of climate change needs to be addressed. As was discussed in Chapter 3, for a long time Russian science was dominated by climate change sceptics. However, after the change happened, the voices actively supporting recognition of the anthropogenic character of climate change and the necessity of its urgent mitigation became more evident and in some cases even went to the other extreme. For example, a journalist of a regional newspaper said, 'I was at one seminar where a host was constantly referring to Al Gore's movie [*An Inconvenient Truth*]. He was convincing us of one point of view, which I did not like. I think they should not adjust facts but present various arguments' (interview 8, Chemal, 13 August 2011).

A climate change reporter for the Russian news agency noted that journalists try to give an opportunity for sceptics to share their point of view in order to write a more balanced (objective) article (interview 30, Moscow, 20 July 2011). The problem of journalists' desire to cover climate change in a balanced manner and to what consequences it leads has been widely discussed in the literature (see Chapter 1). In the Russian case, this problem should be addressed in a more cautious way due to the state's history of climate scepticism and the history of the very open propaganda of the Soviet time, in which case the presentation of one point of view might lead to its total rejection. Hence, this problem could be solved through the improvement of journalists' ability to evaluate scientific controversies.

In conclusion, the importance of the role of experts in this topic cannot be underestimated, and it was stressed on many occasions by journalists themselves. For example, the editor of the regional newspaper said that in order to write about climate change, journalists do not have to become experts themselves, but they just need to have a reasonably good understanding of the issues and know the main trends and the expert community' (interview 16, Chemal, 14 August 2011).

Environmental NGOs as spokesgroups for climate change

Environmental NGOs became one of the best sources of information during the fieldwork conducted for this research project. Due to the specifications of their work, they had vast knowledge of the various aspects of Russia's climate change policy, such as the positions of the state, business and science and, of course, their experience working with journalists. As the experts of the Russian Regional Environmental Centre confirmed, NGOs play a great role in educating people and disseminating information. Furthermore, NGOs see themselves as at the centre of interactions among various actors in climate change policy, 'as a unique keeper of climate information in Russia' (Bogdan et al. 2009). However, as media analysis (Chapter 4) demonstrated, even though NGOs are the third most popular source of information for journalists writing on climate change, they lag quite far behind and overall they are mostly dominated by two influential NGOs: WWF-Russia and Greenpeace Russia.

Nevertheless, environmental NGOs indeed serve as a 'unique' source of information on climate change, or at least on how to find this information. Quite often members of NGOs themselves become, or already were, academics or they integrate the scientific community in their work or collaborate with it on the problem.

Oleg Yanitsky (2009), drawing on 20 years of research on environmental movements in Russia, classifies scientists involved with the work of the NGOs into five categories: neutral, aware, involved, partner and fully integrated. In sum, 'neutral' scientists only provide some expertise for certain projects. The scientists who fall into the 'aware' category, in addition to sharing their knowledge on the subject, also get concerned with the problem. The 'involved' type of scientists remain affiliated with their academic institutions but also share the NGOs' ideas and even take part in their actions. The fourth type, the 'partner', brings collaboration between the scientist and the NGO to the next level, and in this case scientists are officially affiliated with academic institutions but also with the NGO. The final type, 'fully integrated', characterises scientists who no longer work in academia but are fully employed by the NGO.

There are no data on how many scientists are involved with the NGOs working on climate change issues in Russia and at what level. For instance, Podgorny from Greenpeace Russia shared the experience of initiating a Greenpeace project on climate change together with scientists, but they struggled to find scientists who would cooperate. The problem was that when scientists found out that they had to work with Greenpeace, they were very cautious. As Podgorny said, 'they just did not know much about us and thought we only throw paint at fur coats' (interview, Moscow, July 2011). It seems that in the interest of promoting the topic in the media, collaboration between these two groups [scientists and activists] needs to be at least at the third level of Yanitsky's classification. In this case it could be suggested that science and the NGOs might become unified and consequently have a stronger voice in their communication with the media and would illuminate possible problems. There is the possibility of another problem arising – that the merger between the scientific and activist communities might compromise the objectivity of the scientific information.

Another problem which at the moment stops the NGOs from being a stronger voice in the climate change discourse is their ambiguous relations with the state. On the one hand, activists themselves share their negative experiences of trying to communicate with state officials involved in climate change policy (even if it is just to get access to the Russian official delegation at the international conferences on climate change) (various interviews, 2011). On the other hand, state officials acknowledge the existing problems, such as during Medvedev's speech at the conference 'Rio+20'. The former president stated: 'there are around 80 environmental NGOs in Russia [. . .] but of course it is not easy to work with them: environmental organisations are *difficult* partners, but because of this the state needs to support them' (Medvedev 2012).

Perhaps the NGOs are considered '*difficult*' because they are among the few groups of Russia's civil society that constantly question and try to scrutinise the government's performance. As an example, Medvedev's speech at the 'Rio+20' conference and the overall Russian performance there[8] were described as a total failure by the NGO 'Ekologicheskaya Vakhta po Severnomu Kavkazu' [Environmental Watch in the North Caucasus] (2012). On its website, the NGO's activists published an article titled 'At the Summit "Rio+20" Russia Became One of the Countries Unable to Make Environmentally Responsible Decisions.' In the subheading, they go even further by directly insulting the newly appointed prime minister: 'Dmitry Medvedev *lied* to the international community about Russia's successes in the sustainable development sector.'

Unfortunately, sometimes the NGOs themselves act as a barrier to disseminating knowledge on climate change by being resistant to communicating with journalists – 'I often fight or argue with our journalists, because they often call when some disaster happens and ask us what does it mean and what to do, in this case I redirect them to the Ministry for Emergencies and Elimination of Consequences of Natural Disasters, because with us [environmentalists] they should talk about how to prevent these disasters, not what to do when they have already happened' (anonymous source, interview, Moscow, July 2011). The frustration of this environmental activist can be understood, since it seems rather useless in some cases to talk about a disaster which has already happened and when there is nothing that can be done by the NGO to fight its outcomes, whilst at the same time the NGO's messages of warning are ignored by mass media. It could also be seen as a wasted opportunity to attract and maintain journalists' interest in the subject, especially if the topic is connected with global climate change (a prolonged process) which is characterised by the numerous natural disasters and requires constant attention.

In conclusion, the NGOs indeed have great potential for popularising the problem of climate change in Russia and they have been doing so for many years in the forms of special conferences, events, seminars, trips to places where consequences of climate change can be observed and so on. However, there are still multiple problems which to some extent could be solved through enhancing communication links between the NGOs and other actors involved in the process of disseminating information on climate change.[9]

Concluding remarks

In the Russian case of media coverage, the state is clearly at the top of the hierarchical ladder. The state dominates media discourse and plays a significant role in elite consensus. Hence, even though modern mass media in Russia need to operate according to the logic of the free market, before the economic factors 'take off' and start to take part in shaping media messages, the information passes through state elite consensus. This is evident in the case of climate change when the state slightly changed its stance on the problem and finally publicly acknowledged the anthropogenic character of climate change and Russia's commitments to GHG emissions reduction goals – the climate change topic has entered the media discourse with the overarching message of Russia's leading position in the international negotiation process and a 'win-win' situation for the state.

Besides the biased coverage of climate change in Russian newspapers and the absence of any critique of the elites' position on the problem, the clear omission of the topic from media discourse can be seen. 'Climate silence' was demonstrated by the small number of articles published on the subject, through the interviews with people involved in these processes and by the data presented in public opinion polls. It has been argued that the topic was involuntarily removed from the public discourse, since it did not fall into the sphere of elites' interests. Hence, we can talk about the 'third dimension of power', which does not refer to the 'decision making' or 'non–decision making' processes but rather describes the situation where the social system is organised in a way that the interests of the powerful elites are met effortlessly.

In the case of climate change in Russia, the possibilities of altering the media coverage can be presented in the following ways: since open critique of the government is risky and undesirable, the discussion could be brought down to the local level or raised to the international level; the economic benefits of climate change mitigation policies could be popularised. The scientific and activist communities need to become more vocal and find ways of cooperating with each other and other interested parties. Most of these practices have already been taking place for a couple of years; however, in order for Russia to take the discussion of climate change to the next level, they need to be developed further.

Notes

1 The data were collected from 1 December 2009 to 22 December 2009 from two national newspapers, one of which is supposed to 'have a rather close relationship to the local power elite'; and the other is 'a "tabloid" paper or a more consumer-oriented outlet of journalism' (Eide et al. 2010, p. 19).
2 According to Yanitsky (2009), the third phase starts after 2000, whilst the first and the second one occurred in the 1960s–1970s (when only a restricted amount of prominent scientists dominated the environmental discourse in media) and the late 1980s to early 1990s, when any group could have raised the issue (see more in Chapter 3).
3 The same results for the Copenhagen Conference were achieved by the research of the network 'MediaClimate'. For the Russian case the analysis of the 'principal groups of actors quoted' in the newspapers showered that the category 'national political system' was twice as large as the next largest category – 'science, expertise' (Yagodin 2010).
4 For example, members of the EU with a strong pro–climate change mitigation policy.
5 Despite the popular opinion that Russia among other northern countries will agriculturally benefit from the prolonged harvesting season, it has been observed that temperature rise will also result in greater droughts in the most fertile regions of the country, which would counter any benefits of climate change in other regions (Dronin and Kirilenko 2008).
6 For examples, see Davydova 2013, 2014 and Shapovalov 2013.
7 Once again, it should be noted that the positive qualitative and quantitative trend in media coverage of climate change is not understood in the sense that the media should start to publish more articles in favour of taking action towards climate change but maintain its rudimentary level of analysis. Instead the positive change is the one signified by the more sophisticated and knowledgeable understanding of the problem.
8 At the UN Conference on Sustainable Development (Rio+20), which took place from 20 June 2012 to 22 June 2012 in Rio de Janeiro, Brazil (RIA Novosti 2012), Dmitry

Medvedev acknowledged once again the necessity to develop sustainable economic models which would allow for neutralising the environmental threat. He also reported that Russia was successful with its commitments to the Kyoto Protocol and that GHG emissions would be reduced by 25 percent (of the level of 1990) by 2020 and that Russia expected the same from other countries and would participate in a global agreement only if all countries would take part in it (Medvedev 2012).

9   For an extensive study on the environmental NGOs in Russia, consult Laura Henry (2010) and David Feldman and Ivan Blokov (2012), which outline, on the basis of extensive empirical evidence, the major problems that environmental NGOs face in their work and how they cope with them and adjust to the existing political, economic and social environment.

Bibliography

Andonova, L. (2008) 'The Climate Regime and Domestic Politics: the Case of Russia', *Cambridge Review of International Affairs*, 21: 483–504.

Babe, R. (2005) 'Newspaper Discourses on Environment'. In: J. Klaehn (ed.) *Filtering the News. Essays on Herman and Chomsky's Propaganda Model*, London: Black Rose Books: 187–222.

Bachrach, P. and Baratz, M. (1962) 'Two Faces of Power', *American Political Science Review*, 56/4: 947–952.

Bagirov, A. T. and Safonov, G. (2010) *Energobezopasnost' i Klimat: Global'nye Vyzovy dlya Rossii*, Moscow: Teis.

Bogdan, L., Dobrolyubova, Y., Kozel'tsev, M. and Surovikina, E. (2009) *Politika i Deyatel'nost' Rossii v Oblasti Ekologii i Izmeneniya Klimata*, Moscow: Rossiyskiy Regional'nyy Ekologicheskiy Tsentr.

Carroll, C. and McCombs, M. (2003) 'Agenda-Setting Effects of Business News on the Public's Images and Opinions about Major Corporations', *Corporate Reputation Review* 6: 36–46.

Chomsky, N. (1998) 'Propaganda and Control of the Public Mind'. In: R. McChesney, E. M. Wood and J. B. Foster (eds.) *Capitalism and the Information Age: The Political Economy of the Global Communication Revolution*, New York: Monthly Review Press: 179–189.

Climate Doctrine of the Russian Federation (2009), http://archive.kremlin.ru/eng/text/docs/2009/12/223509.shtml, date accessed 10/01/2010.

Curran, J. (2002) *Media and Power*, London: Routledge.

Davydova, A. (2013) 'Zasukhi ne Poluchat Obchego Byudzheta', *Kommersant* (16 November), www.kommersant.ru/doc/2345592?isSearch=True, date accessed 09/08/2014.

Davydova, A. (2014) 'Tsena Izmeneniy Klimata Rastet', *Kommersant* (1 April), www.kommersant.ru/doc/2443235?isSearch=True, (accessed 09/08/2014).

Dronin, N. and Kirilenko, A. (2008) 'Climate Change and Food Stress in Russia: What if the Market Transforms as It Did During the Past Century?' *Climatic Change*, 86: 123–150.

Eide, E., Kunelius, R. and Kumpu, V. (eds.) (2010) *Global Climate – Local Journalisms. A Transnational Study of How Media Make Sense of Climate Summits*, Bochum: Projektverlag.

Ekologicheskaya Vakhta po Severnomu Kavkazu (2012) 'Na Sammite "Rio+20" Rossiya Stala Odnoy iz Stran, ne Sposobnyx k Prinyatiyu Ekologicheski Otvetstvennykh Resheniy', www.ewnc.org/node/9172, date accessed 5/07/2012.

Feldman, D. and Blokov, I. (2012) *The Politics of Environmental Policy in Russia*, Cheltenham: Edward Elgar.

FOM (2008) 'Ekologicheskaya Situatsiya v Massovom Soznanii Rossiyan', www.fom.ru, date accessed 20/09/2011.

Hansen, A. (2010) *Environment, Media and Communication*, London: Routledge.

Hanson, P. (2010) 'Managing the Economy'. In: S. White, R. Sakwa and H. Hale (eds.) *Developments in Russian Politics 7*, Durham, NC: Duke University Press: 188–205.

Henry, L. (2010) *Red to Green: Environmental Activism in Post-Soviet Russia*, Ithaca, NY: Cornell University Press.

Heyward, C. (2007) 'Revisiting the Radical View: Power, Real Interests and the Difficulty of Separating Analysis from Critique', *Politics*, 27/1: 48–54.

Kokhanova, L. (2007) *Ekologicheskaya Zhurnalistika, PR i Reklama*, Moscow: Yuniti-Dana.

Lukes, S. (2005) *Power: A Radical View*, New York: Palgrave Macmillan.

Mandrillon, M.-H. (2008) 'Debating Kyoto: Soviet Networks and New Perplexities'. In: S. White (ed.) *Media, Culture and Society in Putin's Russia*, Basingstoke: Palgrave Macmillan: 133–153.

McCombs, M. (2004) *Setting the Agenda: the Mass Media and Public Opinion*, Cambridge: Polity.

McCombs, M. (2005) 'A Look at Agenda-Setting: Past, Present and Future', *Journalism Studies* 6/4: 543–557.

McCombs, M. and Shaw, D. (1972) 'The Agenda-Setting Function of Mass Media', *Public Opinion Quarterly*, 36/2: 176–187.

Medvedev, D. (2012) 'Rost Ekonomiki i Sberezhenie Prirody – Strategicheskie Zadachi, i Oni Dolzhny Byt' Sbalansirovany', http://blog.da-medvedev.ru/post/235/transcript, date accessed 10/02/2013.

Monaghan, A. (2012) 'The Vertikal: Power and Authority in Russia', *International Affairs*, 88/1: 1–16.

Mountford, S. (2009) 'Who Says Climate Change Matters?', *BBC News* (23 December), www.bbc.co.uk/worldservice/worldagenda/2009/12/091222_worldagenda_copenha gen_poll.shtml, date accessed 08/09/2014.

Nenashev, M. (2010) 'Nezavisimost' SMI – Illyuzii i Real'nost'. In S. Konovchenko (ed.) *Sovremennye SMI Rossii: Teoriya i Praktika. Sbornik Nauchnykh Statey*, Moscow: Moskovskiy Gosudarstvennyi Universitet Pechati: 5–18.

Oates, S. (2007) 'The Neo-Soviet Model of the Media', *Europe-Asia Studies*, 59/8: 1279–1297.

Pralle, S. (2009) 'Agenda-Setting and Climate Change', *Environmental Politics*, 18/5: 781–799.

President of Russia website (2013a) 'Ramkakh Ekologicheskoy Aktsii "Chas Zemli" Budet Vyklyucheno Vneshnee Ocveshchenie Moskovskogo Kremlya', www.kremlin.ru/accreditation/17724, date accessed 30/03/2013.

President of Russia website (2013b) 'Rezul'taty Sotsiologicheskogo Oprosa Naseleniya Rossiyskoy Federatsii po Problemam Izmeneniya Klimata', http://state.kremlin.ru/face/19203, date accessed 08/09/2014.

President of the Russian Federation (10 August 2012) Decree no. 1157 'On Conducting the Year of the Environmental Protection in the Russian Federation'.

President of the Russian Federation (13 December 2012) Decree no. 563-rp 'On the Inter-agency Working Group on Climate Change and Sustainable Development'.

President of the Russian Federation (30 September 2013) Decree no. 752 'On Reduction of Greenhouse Gas Emissions'.

RIA Novosti (2012) 'Medvedev Vstretitsya s Gensekom OON v Rio-de-Zhaneyro' (19 June), http://ria.ru/politics/20120619/676845263.html, date accessed 25/06/2012.

Roshydromet (2012) *Izmeneniya Klimata*, 30, http://global-climate-change.ru/downl/byulletenyo/Izmenenie_klimata_N30_January_2012.pdf, date accessed 14/02/2013.

Roshydromet (2014) *Second Assessment Report of Climate Change and Its Consequences in the Russian Federation.* Moscow: Roshydromet.

Rossiyskaya gazeta (2013) 'Ukaz Prezidenta Rossiyskoy Federatsii ot 30 Sentyabrya 2013 g. N 752', *Rossiyskaya gazeta* (4 October), www.rg.ru/2013/10/04/eco-dok.html, date accessed 05/09/2014.

RSEU [Russian Social-Ecological Union] (2013) 'Bol'shinstvo Rossiyan Znayut ob Izmenenii Klimata', www.rusecounion.ru/klimat_30913, date accessed 08/09/2014.

Schmidt, A., Ivanova, A. and Schafer, M. (2013) 'Media Attention for Climate Change around the World: A Comparative Analysis of Newspaper Coverage in 27 Countries', *Global Environmental Change*, 23: 1233–1248.

Shapovalov, A. (2013) 'IFC Otsenila Kommercheskiy Potentsial Klimaticheskikh Investitsiy', *Kommersant* (6 November), www.kommersant.ru/doc/2336800?isSearch=True, date accessed 09/08/2014.

Smith, N. and Leiserowitz, A., (2013). 'American Evangelicals and Global Warming', *Global Environmental Change*, 23: 1009–1017.

Stern, N. (2007) *The Economics of Climate Change – The Stern Review*, Cambridge: Cambridge University Press.

Tarusin, M. and Fedorov, V. (2009) 'Vzaimootnosheniya Gosudarstva i Sredstv Massovoy Informatsii. Pretenzii Naseleniya k SMI', *Monitoring Obshchestvennogo Mneniya*, 5/93.

White, S. and Oates, S. (2003) 'Politics and the Media in Post-Communist Russia', *Politics*, 23/1: 31–37.

World Bank (2010) 'Public Attitudes toward Climate Change: Findings from a Multi-Country Poll', *The World Development Report*, July.

Yagodin, D. (2010) 'Russia: Listening to the Wind – Clientelism and Climate Change'. In: E. Eide, R. Kunelius and V. Kumpu (eds.) *Global Climate – Local Journalisms. A Transnational Study of How Media Make Sense of Climate Summits*, Bochum: Projektverlag: 275–290.

Yanitsky, O. (2009) 'The Shift of Environmental Debates in Russia', *Current Sociology*, 57/6: 747–766.

Zassoursky, I. (2004) *Media and Power in Post-Soviet Russia*, New York: M. E. Sharpe.

Zhao, X. (2009) 'Media Use and Global Warming Perceptions. A Snapshot of the Reinforcing Spirals', *Communication Research*, 36/5: 698–723.

6 Conclusion

For almost 70 years in Russia, political, economic and social life was determined by the communist nature of the state, where a centrally planned economy was managed by a one-party government, whilst the mass media and other social institutions served the interests of the ruling elite by defending and propagandising the ideals of communist society. While the methods of building and sustaining this political regime were openly criticised by the West and sometimes silently questioned by Russian citizens, in general hardly anyone would be surprised to find out that *Pravda* would not publish journalists' investigations on issues such as the CPSU's budget policy or that western Europe provided better welfare than the USSR.

After the fall of the Soviet Union, it was expected that Russia would move to become a liberal democratic state; however, instead of this, the state fell into a 'grey area'. Russia still qualifies as a democracy, but with a 'managed' or 'illiberal' nature. It is a capitalist state, but the scale of state interference and the level of corruption often intercept the invisible hand of the market. The protection of human rights is not always ensured due to the faults in the state's legal system. Finally, the long-awaited freedom of speech, which at the birth of the new state was recorded in the Russian Constitution, is questioned on a regular basis, especially given the murder of prominent journalists, the threatening of their lives, forceful closure of media outlets and financial sanctions towards the 'fourth estate'. In the modern Russia, mass media have also become participants in the free market economy and have to adjust to market mechanisms. However, it is not always clear when the media are guided by external actors and are being censored or suppressed or when they are just following the logic of the free market and fitting within the new capitalist society. These processes become even more unclear when we look at the news production process on the controversial topic of climate change.

Climate change is one of the biggest and most ambiguous challenges for the world, bringing together science, politics, economics and people who are directly influenced by its consequences. In Russia the climate change topic becomes even more complicated, since according to the most recent data the country is extremely vulnerable to the effects of climate change but also finds itself amongst the largest GHG emitters in the world, which occasionally has made it one of the key figures during the international climate change mitigation process. Russia 'earned' its high status of being one of the biggest contributors to climate change due to the fact that its economy heavily relies on the extraction and export of fossil fuels and in general the Russian economy is extremely carbon intensive (due to climate conditions and the obsolescence of its infrastructure). To make the case even worse, the Russian energy industry is closely connected with the state, to the extent that it becomes virtually impossible to separate the state from business interests where climate policy is concerned.

Furthermore, Russia inherited the 'Soviet legacy' of neglecting environmental problems and sacrificing the environment to economic development. In the late 1980s and early 1990s for a short period of time environmental issues moved up on the Soviet/Russian agenda; however, after two decades of major political transformation which included economic decay and social destabilisation, they were pushed down again. For a long time (and still to this day) economic development became the priority, pushing all other problems, and especially environmental ones, to the background, leading to the policy of 'de-environmentalism'. All of these factors are important because it has been argued that media coverage is directed by the consensus achieved amongst the political and economic elites of the country. To predict the direction of media coverage of climate change in Russia, we need to know the priorities of the main actors involved in the climate policy.

The analysis presented in this study has demonstrated that the Russian media production process is heavily influenced by external actors. The state and big industries (with close connections to the state) own major media outlets in Russia. They dominate the advertising market, and Russian journalists often rely on official sources in their pursuit of information. The state often becomes the monopolist in the production of 'flak', or negative reaction to media messages, and finally, as already mentioned, the regime or dominant ideology of Russia is classified as an 'illiberal' democracy where powers often get abused. In this sense it can be argued that media production processes in Russia are influenced and even managed by the consensus among elite groups. However, the analysis of media coverage of climate change in Russia has demonstrated that in this particular case, the situation is not that drastic. Journalists are not killed for writing about GHG emissions, newspapers do not have quotas on how many articles they should produce on climate change and Russia's climate policy is not classified information. Furthermore, the coverage is not significantly altered depending on differences in newspapers' ownership structures or advertising policies. At the same time, we can see how the media follow the state's position on climate change policy – this is particularly obvious when considering the analysis of the 'sources of information' filter, which has been dominated by official Russian sources (especially after the change in state policy towards becoming more proactive in climate change mitigation).

The correlation between the state's climate policy and journalists' interests towards the topic was also reaffirmed during the interviews with experts working in the media or with the media. Indeed, until a few years ago the problems were not on the agenda of the Russian government and were addressed only at international negotiations in order to receive some benefits from external actors, whilst at the domestic level climate change was often ridiculed and its anthropogenic character was not accepted. At the same time, media coverage and media interest towards climate change issues were practically non-existent. It has been argued that the state has succeeded in removing the issue from the agenda, or in other words, it has successfully exercised its 'third dimension of power'. Thus, it did not force journalists to write or not to write about the problem, but since the state acts as a 'main newsmaker' on the topic and due to its reluctant attitude towards climate change, this issue did not even enter public discourse.

A few years ago (around the time of the Copenhagen Conference in 2009), attention to the problem had risen on both sides – the state and the media. The most important change was the official acknowledgement of climate change and its anthropogenic origins and the announcement of Russia's commitments to GHG emissions reduction goals. Arguably, these changes in state policy were motivated by the realisation of the economic benefits of following a low emissions policy. Russia has a great capacity to cut its carbon contribution to climate change and at the same time keep developing and modernising its economy. This approach has been described as 'climate pragmatism', where the environment is still assessed from an economic point of view, but this time it leads to a pro-environment oriented economy. In the media the change was signified by an increase in coverage on climate change and the appearance of journalists specialising on the topic – which once again brings us to the idea that in this case the media are not openly managed by the ruling elite, but rather the 'manufacturing of consent' happens unintentionally. Journalists willingly follow the rules of the game, or in our case they follow the lead of their most important 'source of information' without any specific orders from the top: 'journalists are not paid by somebody to report climate change news in the way they do (that climate change is made by space aliens or we are all doomed), but they do it because they write what they are interested in [or because it sells]. It is not a political order. And to be honest if something indeed was "ordered", it would be impossible to prove it (as happened in the case of Andrey Illarionov[1])' (environmental activist, interview, Moscow, 27 July 2011).

Coming from this proposition, it was suggested that in the particular case of climate change coverage in Russia, the state elites serve as the initial barrier or even motivators or de-motivators of the journalists' interest in the topic. Further on, economic factors play a part in shaping media discourse on the subject (journalists still need to consider how to sell their newspapers and also not to create unnecessary conflicts with their owners or advertisers, even though these factors matter to a lesser degree than state elites' interest in the topic). If the topic falls in the sphere of interests of these elite groups, then the micro-factors start to play their part and journalists begin to think about how they have to deal with the problem. Therefore, the changes described in the media coverage of climate change in Russia should not be simplified to the straightforward relations between the state and the media outlets. The journalists did mention that climate change as a topic dictates its own rules – it requires specialised knowledge, it has low appeal to the public as being a prolonged and intangible process, it takes extra time to find the right information sources and so on.

In conclusion, the Russian media system still is in a state of great dependency on the government and often has to play by its rules and adjust its content, or pay the price. In the case of climate change, the media are not 'ordered' to write about environmental change, but rather they just follow the interests of the 'main news-maker'. Naturally, this seems to be a negative situation, where the media do not question state performance (such as its extremely weak domestic policy on climate change) but rather restate the elites' vision of the problem that Russia has great potential to become an environmental donor and, as Medvedev claimed before the Copenhagen Conference, 'we will win, no matter what'. As previous studies on media and climate change have shown, sometimes unrestrained media cover-age of climate change can be counterproductive. There is the possibility of the abuse of alarmist messages, controversy, the tendency to cover the problem in a supposedly balanced way by providing too much space to climate sceptics and so on (even though they are very much a minority in the scientific community). The Russian case demonstrates that following a single actor is likely to result in greater orthodoxy – the coverage becomes more coherent, but does it make it more adequate? As one of the interviewees (journalist, interview 2, Moscow, 22 July 2011) put it: 'our main official TV channel reproduces the state's messages, which is not always bad. Many of Medvedev and Putin's decisions are good, they also understand that Russia needs to develop. But, in my opinion, they should some-times step aside'. Naturally this tendency of the media following the state's lead on climate issues is somewhat dangerous, since if the state policy will again become even more reluctant, then the topic will disappear from the agenda as well.

Climate change is a problem which will not be solved in a few years' time, and we will eventually understand this fact more clearly, as will the Russian govern-ment. Regardless of whether the Russian state will try to modernise its economy and attract more investments by making it more sustainable, positioning itself as an 'environmental leader' due to its vast reserves of natural resources and poten-tial for carbon sinking (through its forests) or trying to reduce economic loss from the consequences of climate change due to its great vulnerability, there is an extremely small chance that the topic will disappear from its agenda and therefore from media discourse. Hence, it is up to civil society and the journalists them-selves to find the loopholes in the system to diversify and improve climate change coverage.

Note

1 Illarionov is famous for promoting his climate change sceptic position and for years actively lobbied against the ratification of the Kyoto Protocol. Some argued that his point of view was paid for by Western hydrocarbon companies (see more in Chapter 3).

Appendix

Overview of interviewees

Interview number	Name	Position/affiliation	Date	Location of the interview
1	Anonymous source	Municipal administration of a regional capital	27 May 2011	Skype interview
2	Anonymous source	Former news editor of national TV channel	22 July 2011	Moscow
3	Belova, Valentina	Specialist in State Environmental Expertise, Department of Nature and Environmental Protection in Altay Region	10 August 2011	Barnaul
4	Berdin, Vladimir	Head of the Department of Sustainable Development and Partnership at the Sustainable Energy Development Centre	13 August 2011	Chemal
5	Anonymous source	Correspondent of the Russian branch of an international broadcaster	25 July 2011	Moscow
6	Anonymous source	Environmental NGO	27 July 2011	Moscow
7	Anonymous source	Correspondent, Russian national newspaper	7 July 2011	Skype interview
8	Anonymous source	Correspondent, regional newspaper	13 August 2011	Chemal
9	Dobrovidova, Olga	Special climate correspondent, news agency RIA Novosti	22 July 2011	Moscow
10	Dr. Dobrynin, Dmitriy	Moscow State University	14 August 2011	Chemal
11	Harrison, John	Climate change host, Voice of Russia (radio)	18 June 2012	Skype interview
12	Kharitonov Leonid	Head of the Department of Nature and Environmental Protection in Altay Region	10 August 2011	Barnaul

(*Continued*)

Interview number	Name	Position/affiliation	Date	Location of the interview
13	Dr. Kharlamova, Nataliya	Associate professor, geography faculty, Altay State University	February 2012	Email
14	Kokorin, Alexey (PhD)	Climate change programme coordinator, WWF-Russia	27 July 2011	Moscow
15	Anonymous source	Environmental NGO	July 2011	Email
16	Anonymous source	Editor, regional newspaper	14 August 2011	Chemal
17	Anonymous source	Correspondent, host of a radio programme for a regional radio station	13 August 2011	Chemal
18	Anonymous source	Environmental NGO	14 July 2011	Skype interview
19	Anonymous source	Environmental NGO	27 July 2011	Moscow
20	Dr. Safonov, Georgiy	Director, NGO 'Centre of Environmental Innovations'	13–14 August 2011	Seminar presentation, Chemal
21	Anonymous source	Journalist, TV news	13 August 2011	Chemal
22	Dr. Stetsenko, Andrey	NGO 'Centre of Environmental Innovations'	14 August 2011	Chemal
23	Surovikina, Elena	PR consultant on environmental issues, United Nations Development Programme in Russia	14 August 2011	Chemal
24	Uporova, Elena	Deputy editor-in-chief of 'Fond Nezavisimogo Radioveshchaniya' (Fund for Independent Radio)	14 August 2011	Seminar presentation, Chemal
25	Dr. Vargin, Pavel	Research Scientist in Laboratory of Experimental Middle Atmosphere Research of Central Aerological Observatory, editor of the newsletter *Izmenenie Klimata* ['Climate Change']	April 2013	Email
26	Anonymous source	Environmental NGO	9 August 2011	Barnaul
27	Anonymous source	Environmental NGO	22 July 2011	Moscow

(*Continued*)

Interview number	Name	Position/affiliation	Date	Location of the interview
28	Anonymous source	Editor, regional branch of a national news agency	18 August 2011	Barnaul
29	Zakharov, Vladimir (PhD)	Director, NGO 'Centre for Environmental Policy and Culture'	21 July 2011	Moscow
30	Anonymous source	Correspondent, national news agency	22 July 2011	Moscow
31	Anonymous source	Environmental NGO	27 July 2011	Moscow

Index